Study Guic

for

Sociology

A Global Perspective

Study Guide

for

Sociology

A Global Perspective

Sixth Edition

Joan Ferrante
Northern Kentucky University

Australia • Brazil • Canada • Mexico • Singapore • Spain • United Kingdom • United States

Printed in Canada

1 2 3 4 5 6 7 09 08 07 06 05

Printer: Webcom

ISBN 0-495-00563-0

Flip Chalfant/Getty Images

Thomson Higher Education
10 Davis Drive
Belmont, CA 94002-3098
USA

Contents

Preface

I wrote the *Study Guide for Sociology: A Global Perspective* 6th edition hoping to make it more than simply a test preparation tool. Everyone intuitively knows that memorizing material for the sake of doing well on a test cannot lead to meaningful and long-lasting learning experiences. In other words, if students read the text only with an eye for predicting possible test questions and take class notes without thinking about them until the night before the test, the academic experience will be an empty one. Real learning means thinking about the ideas that you hear and read, taking an active role in learning, and incorporating learning experiences into the activities of daily life. I wrote the Study Guide with this vision of learning in mind. Each chapter corresponds to a chapter in the textbook and includes seven sections: (1) study questions, (2) key concepts, (3) concept application, (4) practice multiple-choice and true/false questions, (5) applied research, (6) internet resources, and (7) statistical profiles. In addition, please visit the supporting Web site at http://sociology.wadsworth.com. Just click on the book cover icon under the Introductory Sociology heading.

Study Questions

Think of the study questions as note-taking tools. If you answer these questions thoughtfully and conscientiously, you will come away with a thorough synthesis of the sociological material covered in the textbook.

Key Concepts

The key terms and concepts from each chapter of the text are listed here, organized in outline format and with page references to where they can be found in the main text, to aid you in reviewing the main ideas of each chapter.

Concept Applications

Each chapter contains several concept application scenarios. You are asked to determine which concept(s) is best represented in that scenario and to explain why.

Practice Multiple-Choice and True/False Questions

If you treat these questions as a comprehensive study tool, you will not be adequately prepared to take a test. On the other hand, if you prepare beforehand (as if you are going to take a test that will count for a grade), the score you earn will probably be a good indicator of how well you can expect to do on the actual test.

Applied Research

The applied research exercises take learning beyond the book and ask students to become learners that are actively involved in gathering information to answer questions that relate to material covered in the textbook.

Internet Resources

For each chapter there are recommended internet resources related to the chapter topic and to the country emphasized in that chapter.

Country Statistical Profile

The country profile are excerpted from *The World Factbook 2005* and/or *Country Background Notes* published by the U.S Central Intelligence Agency and U.S. Department of State. The reports give additional background on various countries or regions of the world emphasized in the textbook.

Chapter 1

The Sociological Imagination

Study Questions

1. In the classic book, *Invitation to Sociology*, Peter L. Berger presents sociology as a form of consciousness. Explain.

2. Sociologists focus on human interaction with emphasis on two key concepts: society and social interaction. Define each.

- society: a system of social interaction
- soccal interactions: everyday events in which the people invowed take one another into account by consciously / unconsciously attaching meaning to the situation, interpretting what others are saying and then responding accordingly.

3. Durkheim maintains that the sociologist's task is to study social facts. What are social facts?

ideas, feelings, ways of behaving "that possess the remarkable property of existing outside the consciousness of an individual.

4. Peter Berger maintains that a "debunking motif" defines the sociological consciousness. Explain.

- debunking: looking beyond the obvious for explanations of human behaviour
- sociologist are driven to debunk the social systems they study

5. Distinguish between troubles and issues.

- Troubles: personal needs, problems, difficulties, explained in terms of the individual shortcomings (motivation, attitude, character, ability)
- issue: matter explained only by factors outside an individuals control and immediate environment.

6. Explain the connection between troubles and institutional crisis.

- most people are unaware of the connection between troubles and institutional crisis.
- ex troubled by hair growth → connected to a capitalistic system that commercializes gender ideals, uses insecurity to sell products.

7. What is the sociological imagination?

- The ability to connect seemingly impersonal/remote historical forces to the most basic incidents of an individuals life.
- enables people to distinguish between personal troubles and public issues.

8. What major historical event shaped the discipline of sociology? Why?

The Industrial Revolution
- interdependent world emerged
- colonizing countries
- innovations: oil, steam, transportation

9. How did the Industrial Revolution affect the nature of work and social interaction?

- goods went from hand made and unique to mass quantaties of machine made goods.
- workshops -> factories, craftsmen -> machines

10. For what writings is Marx most famous?

"Das Kapital" and "The Communist Manifesto"
- critical of capital system
- predicts it's defeat to socialism
- describes capitalism
-

11. Define Karl Marx's vision of the sociologist's task. What concepts and assumptions drive his analysis of society?

- to analyze/explain conflict, the major force that drives social change.
- means of production
- class conflict propelled people from one historical epoch to another.
- causes/consequences of inequality
- bourgeoisie vs. proletariat

12. According to Durkheim, in what way did industrialization affect society?

- changed the nature of the ties that bound people to one another.
- sociologists task - analyze/explain solidarity (essential concern)

13. How did Durkheim explain differences in suicide rates?

- there is no situation that could serve as an occasion for someones suicide.

14. Distinguish between egoistic, altruistic, anomic, and fatalistic suicide.

egoistic - ties attaching individual to society is weak
altruistic - individuals have no life of their own, strive to blend with others to have a sense of being.
anomic - dramatic changes in economic circumstances
fatalistic - no hope of change, no release.

15. What is rationalization?

16. Give an example of action/behavior driven by tradition and compare it with value-rational action.

Traditional - go to college because rest of family has
value-rational - go to college because employers value and demand a diploma
Affectional - go to college for love of learning
instrumental - considers options, choose a goal.

17. Who is Harriet Martineau? How did she conduct research for *Society in America*? What methods did she use to make sense of her observations?

- 1825 englishwomen, listened to casual conversations of all kinds of people
- non-biased, made no judgement
- asked readers to test state of affairs against ideal standard
- judge for themselves.

18. Explain the phrase "strange meaning of being black." What life experience may have influenced DuBois' preoccupation with this phrase?

- double consciousness - sense of always looking at oneself through the eyes of others (an American, a Negro)
- two warring ideas in one dark body.
- father was a Haitian of French/African descent
- mother was an American of Dutch/African descent

19. Why are the ideas of Marx, Durkheim, Weber, Martineau, and Du Bois relevant to understanding today's problems and issues?

- They were there to witness the changes of industrial revolution, we have always known it to be this way.
- observations are still relevant today.

20. Describe up to three assumptions that underlie the global perspective.

- lives of people around the world are intertwined, social relationships don't stop at national borders
- globalization is not new, just changed
- movement of goods/services not a one-way process
- globalization plays out differently depending where you are.

21. Imagine that you majored in sociology. How would you explain the usefulness of the sociological perspective?

Key Concepts

Society	pg. 6
Social interaction	pg. 6
System	pg. 6
Social facts	pg. 7
Sociological imagination	pg. 14
Troubles	pg. 9
Issues	pg. 9
Institution	pg. 13
Conflict	pg. 17
Means of production	pg. 17
Bourgeoisie	pg. 17
Proletariat	pg. 17
Solidarity	pg. 18
Suicide	pg. 19
Egoistic	pg. 19
Altruistic	pg. 19
Anomic	pg. 19
Fatalistic	pg. 19
Social Actions	pg. 20
Traditional	pg. 21
Affectional	pg. 21
Value-rational	pg. 21
Instrumental	pg. 21
Disenchantment	pg. 22
Double consciousness	pg. 23
Globalization from above	pg. 26
Globalization from below	pg. 26

Concept Application

Consider the concepts listed below. Match one or more of the concepts with each scenario. Explain your choices.

- ✓ Bourgeoisie
- ✓ Double consciousness
- ✓ Global interdependence
- ✓ Means of production
- ✓ Troubles/Issues

Scenario 1: "In the end, it wasn't Alzheimer's disease that took the life of retired Baltimore County Circuit Judge Robert I.H. Hammerman. It was his obsession with dying from it. The pages of Hammerman's 10-page suicide note—one he copied and mailed to 2,200 friends and family members—vibrantly illustrate his fear of the disease. The well-respected judge wrote about his mother, who lived to 82 but who died after her Alzheimer's 'reached its final stage.'" (Brune 2004)

trouble

Scenario 2: "The world's trade in bananas is dominated by just three huge food multinationals: United Brands (with a 34 percent market share in 1974), Standard Fruit (with a 23 percent) and Del Monte (10 percent). As with many other commodities, the companies control the transport, packaging, shipment, storage and marketing of the fruit. As a result, the profits from bananas go largely into western pockets, while the producer countries get only a pittance" (Harrison 1987:348).

bourgeoisie

Scenario 3: *Black Soldiers in Jim Crow Texas* introduces readers to African American soldiers who were assigned to one of four black regiments (9th and 10th Cavalries, and 24th and 25th Infantries). Not only did these men bear arms and fight gallantly in the Spanish-American War, but at times they used their military weapons in struggles for racial equality in the United States as well. More than three decades after the Emancipation Proclamation, black soldiers grew intolerant of "racial slurs, refusal of service at some businesses, and harassment." Texas's "lower-status Hispanics, the bulk of the population…shared southern white prejudice against blacks. . .The United States bestowed six Medals of Honor and twenty-six Certificates of Merit on their members, and all four regiments inspired laudatory press coverage." Yet these men faced the indignities of racism when serving at military installations in the United States. (Moore 1996:478)

double consciousness

Scenario 4: Afghanistan is now in its fourth year of drought. Haunted by want, depleted from hunger, Akhtar Muhammad first sold off his few farm animals and then, as the months passed, bartered away the family's threadbare rugs and its metal cooking utensils and even some of the wooden beams that held up the hard-packed roof of his overcrowded hovel. (Bearak 2002:A1)

issue

Scenario 5: Five foreign-born players appeared in the National Basketball Association All-Star Game last month, and another five played in the Rookie Challenge game. Of the 348 active players in the NBA, 49 are from abroad...Lenny Wilken, coach of the Toronto Rapters, has said, "I wouldn't be surprised if there are double the number of players in the next five years or so." (Shields 2002:56)

Applied Research

Find a newspaper or magazine article in which the reporter highlights a seemingly personal problem. Briefly describe the problem. Does the article suggest a cause of the problem? Does the article connect the individual trouble to a larger issue or to flaws or breakdowns in institutional arrangements? If yes, explain. If no, can the problem be explained in terms of a larger issue? How so?

Practice Test: Multiple-Choice Questions

1. Peter L. Berger equates the sociologist with
   a. a curious observer walking down the neighborhood streets of a large city fascinated with what he or she can not see taking place behind the building walls.
   b. an Internal Revenue Service auditor.
   c. a judge giving instructions to a jury.
   d. a talk show host interviewing guests.

2. Sociologists define society as
   a. those endowed with great prestige or privilege.
   b. a system of social interaction.
   c. a special organization of people who work to achieve a goal.
   d. equivalent to a country or nation of people.

3. When sociologists study social interaction they focus on all *but* which one of the following?
   a. the ways in which people who do *not* know each other manage to interact.
   b. the system guiding social interaction.
   c. the personalities of those involved.
   d. the spoken and unspoken rules guiding that interaction.

4. "Because I refuse to shave under my arms, I have to pay a price. On a personal level, this price was my mother's hostility. On a public level, the price is dealing with the stares of strangers." This statement illustrates
   a. mechanical solidarity.
   b. social relativity.
   c. the power of social facts.
   d. the idea of double-consciousness.

5. A woman writes "I can't be anything but what my skin color tells people I am. I am black because I look black. It does not matter that my family has a complicated biological heritage." She is writing about the power of
   -> a. social facts.
   b. troubles.
   c. the sociological imagination.
   d. rationalization.

6. Which of the following explanations would someone use to explain an issue?
   a. "She had the opportunity but didn't take it."
   b. "He is lazy."
   c. "There is a flaw or breakdown in an institutional arrangement."
   d. "She didn't try very hard in school."

7. Sociologist C. Wright Mills believed that to gain some sense of control over their lives, people need
   a. to keep up with the news.
   b. regular breaks from their hectic schedule.
   c. a quality of mind to help them grasp the interplay between their biographies and institutional arrangements.
   d. to take personal responsibility for their actions.

8. A sociologist views a photo of an American soldier and an Iraqi child bumping fists. The image prompts the sociologist to ask:
   a. What does it mean for the U.S. to occupy/liberate a country where 40 percent of the population is 14 and under?
   b. Does the American soldier have a child of his own?
   c. Is the soldier an occupier or a liberator?
   d. How many American soldiers are stationed in Iraq?

9. The period in history known as the Age of Imperialism (1880-1914)
   a. was one of the most peaceful in modern history.
   b. represents the most rapid colonial expansion in history.
   c. preceded the period in history known as the Industrial Revolution.
   d. corresponds with the Cold War between the United States and the former Soviet Union.

10. The Industrial Revolution transformed the nature of work in which one of the following ways?
    a. Machine production was replaced by hand production.
    b. People now could say, "I made this; this is a unique product of my labor."
    c. Products became standardized, and workers performed specific tasks in the production process.
    d. Artisans' power over the production process increased dramatically.

11. *The Communist Manifesto* begins with the line
    a. "A specter is haunting Europe--the specter of Communism."
    b. "Workers of all countries, unite."
    c. "I am not a Marxist."
    d. "The global economy is restless, anxious, and competitive."

12. With mechanization, the rise of two distinct classes emerged. The one that owns the means of production is called the
    a. proletariat.
    b. bourgeoisie.
    c. socialists.
    d. communists.

13. From a sociological perspective, suicide is
    a. an act of intentionally killing oneself.
    b. the result of personal disappointment and sorrow.
    c. self-hatred actualized.
    d. the severing of relationships.

14. ___________________ suicide occurs when people kill themselves because they have been cast into a lower status.
    a. Egoistic
    b. Altruistic
    c. Anomic
    d. Fatalistic

15. A quilt maker may work years creating a one-of-a-kind object from fabrics saved or purchased and then give it to a special person. Weber would classify the quilt maker's actions as driven by
    a. rationalization.
    b. specialization.
    c. an emotion such as love, loyalty, or revenge.
    d. value-rational motives.

16. W. E. B. DuBois coined the phrase
    a. the ties that bind people to one another.
    b. the "strange meaning of being black."
    c. the means of production.
    d. the course and consequences of social action.

17. If scientists discover how to control the aging mechanisms and human life expectancy increases to 150 years, the category of people best able to give insights about the consequence of this change would be
    a. the early sociologists.
    b. those born after this discovery is made.
    c. those who lived both before and after the discovery.
    d. those born a century or more after the discovery.

For the following questions, use one of these responses to identify the thinker associated with each statement.
    a. Karl Marx
    b. Emile Durkheim
    c. Max Weber
    d. Harriet Martineau
    e. W. E. B. DuBois

18. The sociologist's task is to study social facts.

19. In conducting social research it is important to see a country in all its diversity.

20. "The workers have nothing to lose but their chains; they have a whole world to gain. Workers of all countries unite."

Practice Test: True/False Questions

T F 1. Sociologists maintain that love is a violent, irresistible emotion that strikes someone at random.

T F 2. The rules governing social interaction are analogous to the rules governing games such as baseball or monopoly.

T F 3. From a sociological perspective, high unemployment can be solved by changing the negative attitudes of the unemployed.

T F 4. The changes triggered by the Industrial Revolution are incalculable.

T F 5. Karl Marx should be remembered as a student of capitalism.

T F 6. The concept solidarity drives Durkheim's analysis of suicide.

T F 7. Weber maintained that the sociologist's task is to study social action.

T F 8. U.S. dominance in the world can be observed in its military presence in 50 countries.

T F 9. A brand name product such as the Ford automobile is "most likely made in the U.S.A."

T F 10. By definition, social relationships must stop at national borders.

Internet Resources Related to the Sociological Imagination

- **Sociological Tour Through Cyberspace**
  http://www.trinity.edu/~mkearl/
  Sociologist Michael Kearl at Trinity University is interested in cyberspace's potential "to inform and generate discourse, to truly be a 'theater of ideas'". To demonstrate this potential Kearl has created more than 20 such "theaters," which explore topics of interest to any student of sociology, including marriage and family life, social gerontology, social inequality, gender and society, race and ethnicity, and sociology of death and dying.

Internet Resources Related to Global Interdependence

- **Global Interdependence Initiative**
  http://www.aspeninstitute.org/Programt1.asp?i=70
  The Aspen Institute's Global Independence Initiative is grounded in the belief that "global interdependence requires an approach to international engagement that better balances military, economic, and environmental and humanitarian concerns. Such an approach is more consistent with key American values such as fairness, teamwork, generosity, respect for individual autonomy and democratic participation. A global role guided consistently by these values would serve both the majority of humanity and long-term American interests." The website features several reports including *U.S. in the World: Talking Global Issues with Americans-A Practical Guide* and *From Values to Advocacy: Activating the Public's Support for U.S. Engagement in an Interdependent World.*

- **YaleGlobal On-Line**
  http://yaleglobal.yale.edu/globalization/
  Debate abounds over whether globalization is good or bad for the self, the family, the nation, and the world. Some pessimists see increased interdependence as a terribly destructive trend, while optimists see a more diverse, better life for all. Some people argue that the world is no more globalized than it was in the waning days of the British Empire, but some see an information revolution that is unparalleled in history and widespread in its implications. YaleGlobal On-Line has assembled a series of articles to shed light on this debate. Examples include "Grassroots versus Globalism," "Globalization's Missing Middle," and "Globalization: Europe's Wary Embrace."

Statistical Profile of the World

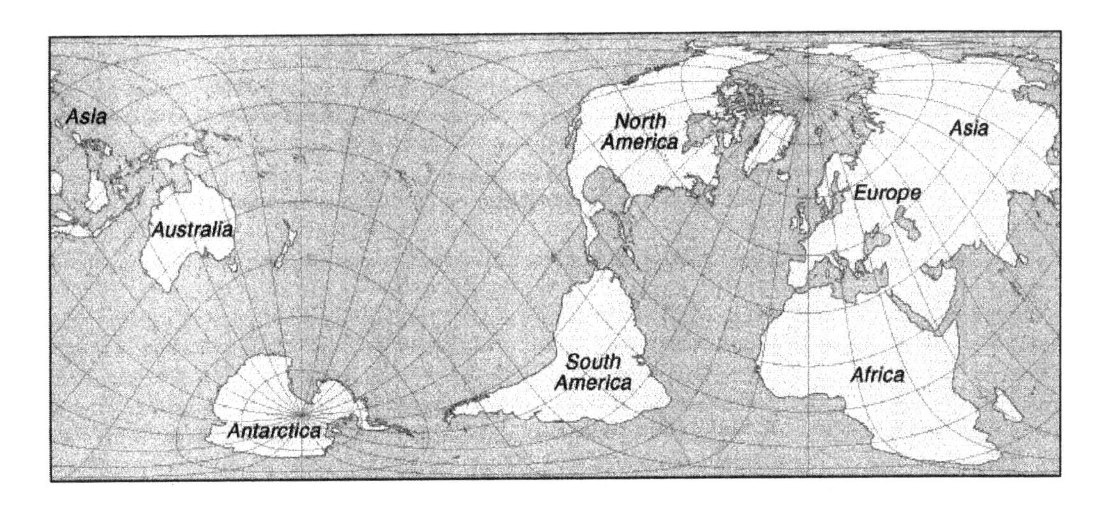

Background

Globally, the 20th century was marked by: (a) two devastating world wars; (b) the Great Depression of the 1930s; (c) the end of vast colonial empires; (d) rapid advances in science and technology, from the first airplane flight at Kitty Hawk, North Carolina (US) to the landing on the moon; (e) the Cold War between the Western alliance and the Warsaw Pact nations; (f) a sharp rise in living standards in North America, Europe, and Japan; (g) increased concerns about the environment, including loss of forests, shortages of energy and water, the decline in biological diversity, and air pollution; (h) the onset of the AIDS epidemic; and (i) the ultimate emergence of the US as the only world superpower. The planet's population continues to explode: from 1 billion in 1820, to 2 billion in 1930, 3 billion in 1960, 4 billion in 1974, 5 billion in 1988, and 6 billion in 2000. For the 21st century, the continued exponential growth in science and technology raises both hopes (e.g., advances in medicine) and fears (e.g., development of even more lethal weapons of war).

Area

Total area: 510.072 million sq km
Land area: 148.94 million sq km
Water area: 361.132 million sq km
Note: 70.8% of the world's surface is water, 29.2% is land
Natural resources: The rapid depletion of nonrenewable mineral resources, the depletion of forest areas and wetlands, the extinction of animal and plant species, and the deterioration in air and water quality (especially in Eastern Europe, the former USSR, and China) pose serious long-term problems that governments and peoples are only beginning to address.

Environment

Current issues: Large areas subject to overpopulation, industrial disasters, pollution (air, water, acid rain, toxic substances), loss of vegetation (overgrazing, deforestation, desertification), loss of wildlife, soil degradation, soil depletion, erosion.
Geography note: The world is now thought to be about 4.55 billion years old, just about one-third of the 13-billion-year age estimated for the universe.

People

Population:	6,379,157,361 (July 2004 est.)
Age structure:	*0-14 years:* 28.2% (male 925,276,767; female 875,567,830) *15-64 years:* 64.5% (male 2,083,789,165; female 2,033,226,759) *65 years and over:* 7.2% (male 203,286,504; female 257,705,851) *note:* some countries do not maintain age structure information, thus a slight discrepancy exists between the total world population and the total for world age structure (2004 est.)
Religions:	Christians 32.71% (of which Roman Catholics 17.28%, Protestants 5.61%, Orthodox 3.49%, Anglicans 1.31%), Muslims 19.67%, Hindus 13.28%, Buddhists 5.84%, Sikhs 0.38%, Jews 0.23%, other religions 13.05%, non-religious 12.43%, atheists 2.41% (2002 est.)
Languages:	Chinese, Mandarin 14.37%, Hindi 6.02%, English 5.61%, Spanish 5.59%, Bengali 3.4%, Portuguese 2.63%, Russian 2.75%, Japanese 2.06%, German, Standard 1.64%, Korean 1.28%, French 1.27% (2000 est.) *note:* percents are for "first language" speakers only

Economy

Global output rose by 3.7% in 2003, led by China (9.1%), India (7.6%), and Russia (7.3%). The other 14 successor nations of the USSR and the other old Warsaw Pact nations again experienced widely divergent growth rates; the three Baltic nations continued as strong performers, in the 5%-7% range of growth. Growth results posted by the major industrial countries varied from a loss by Germany (-0.1%) to a strong gain by the United States (3.1%). The developing nations also varied in their growth results, with many countries facing population increases that erode gains in output. Externally, the nation-state, as a bedrock economic-political institution, is steadily losing control over international flows of people, goods, funds, and technology. Internally, the central government often finds its control over resources slipping as separatist regional movements - typically based on ethnicity - gain momentum, e.g., in many of the successor states of the former Soviet Union, in the former Yugoslavia, in India, in Iraq, in Indonesia, and in Canada. Externally, the central government is losing decision-making powers to international bodies. In Western Europe, governments face the difficult political problem of channeling resources away from welfare programs in order to increase investment and strengthen incentives to seek employment. The addition of 80 million people each year to an already overcrowded globe is exacerbating the problems of pollution, desertification, underemployment, epidemics, and famine.

Because of their own internal problems and priorities, the industrialized countries devote insufficient resources to deal effectively with the poorer areas of the world, which, at least from the economic point of view, are becoming further marginalized. The introduction of the euro as the common currency of much of Western Europe in January 1999, while paving the way for an integrated economic powerhouse, poses economic risks because of varying levels of income and cultural and political differences among the participating nations. The terrorist attacks on the US on 11 September 2001 accentuate a further growing risk to global prosperity, illustrated, for example, by the reallocation of resources away from investment to anti-terrorist programs. The opening of war in March 2003 between a US-led coalition and Iraq added new uncertainties to global economic prospects. After the coalition victory, the complex political difficulties and the high economic cost of establishing domestic order in Iraq became major global problems that continue into 2004.

International Disputes

Stretching over 250,000 km, the world's 322 international land boundaries separate the 192 independent states and 70 dependencies, areas of special sovereignty, and other miscellaneous entities; ethnicity, culture, race, religion, and language have divided states into separate political entities as much as history, physical terrain, political fiat, or conquest, resulting in sometimes arbitrary and imposed boundaries. Maritime states have claimed limits and have so far established over 130 maritime boundaries and joint development zones to allocate ocean resources and to provide for national security at sea. Boundary, borderland/resource, and territorial disputes vary in intensity from managed or dormant to violent or militarized; most disputes over the alignment of political boundaries are confined to short segments and are today less common and less hostile than borderland, resource, and territorial disputes. Undemarcated, indefinite, porous, and unmanaged boundaries, however, encourage illegal cross-border activities, uncontrolled migration, and confrontation. Territorial disputes may evolve from historical and/or cultural claims, or they may be brought on by resource competition. Ethnic clashes continue to be responsible for much of the territorial fragmentation around the world. Disputes over islands at sea or in rivers frequently form the source of territorial and boundary conflict. Other sources of contention include access to water and mineral (especially petroleum) resources, fisheries, and arable land. Nonetheless, most nations cooperate to clarify their international boundaries and to resolve territorial and resource disputes peacefully. Regional discord directly affects the sustenance and welfare of local populations, often leaving the world community to cope with resultant refugees, hunger, disease, impoverishment, deforestation, and desertification.

Source: The World Factbook page on the World (2004)
http://www.cia.gov/cia/publications

Chapter References

Bearak, Barry. 2002. "Children as Barter in a Famished Land." *The New York Times* (March 8):A1.

Brune, Adrian. 2004. "Baltimore Judge's Suicide Leaves Friends Puzzled." *Washington Blade* www.washingtonblade.com

Harrison, Paul. 1988. *Inside the Third World: The Anatomy of Poverty*, 2nd ed. New York. Viking Penguin.

Moore, Brenda L. 1996. Review of *Black Soldiers in Jim Crow Texas 1899-1917*, by Garna L. Christian. *Contemporary Sociology* 25(4):478-79.

Shields, David. 2002. "Foreign Guys Can Shoot." *The New York Times Magazine* (March 3): 56-57.

Answers

Concept Application

1. Fatalistic pg. 19
2. Means of production pg. 17
3. Double consciousness pg. 23
4. Troubles/Issues pg. 14
5. Global interdependence pg. 25

Multiple-Choice

1. a	pg. 4		11. a	pg. 17
2. b	pg. 6		12. b	pg. 17
3. c	pg. 6		13. d	pg. 19
4. c	pg. 8		14. c	pg. 19
5. a	pg. 7		15. c	pg. 21
6. c	pg. 14		16. b	pg. 23
7. c	pg. 14		17. c	pg. 24
8. a	pg. 11		18. b	pg. 7
9. b	pg. 15		19. d	pg. 23
10. c	pg. 16		20. a	pg. 17

True/False

1. F	pg. 4
2. T	pg. 6
3. F	pg. 9
4. T	pg. 16
5. T	pg. 17
6. T	pg. 18
7. T	pg. 20
8. F	pg. 25
9. F	pg. 26
10. F	pg. 25

Chapter 2

Theoretical Perspectives and Methods of Social Research
With Emphasis on Mexico

Study Questions

1. Why is Mexico the focus of Chapter 2?

- shows dependence on maquiladoras
- low wage mexican labour

2. How do the three theoretical perspectives help us to think about a particular event (such as U.S. *maquiladora* operations in Mexico)?

- collect, analyze, interpret facts
- apply theory and research

3. What is a function? Give an example.

- contribution of a part to the order and stability within a larger system.

ex: sports team – function is to draw crowds from different backgrounds, loyalty...

4. According to the functionalist perspective, why has poverty not been eliminated?

poverty has a function in society

examples – poor take on menial work, low pay, dangerous jobs
– poor volunteer for drug tests ect
– cops, social workers ect exist to monitor the poor
– poor purchase goods ex: day old bread.

5. What are the major shortcomings of the functionalist perspective?

- is conservative in that it defends existing arrangements
- some parts don't have a function

6. What concepts did Robert K. Merton introduce to counter criticisms of the functionalist perspective? Briefly define each concept and explain how they strengthen the perspective. What criticism is not addressed by Merton's concepts?

manifest functions – intended/anticipated effects on order/stability
latent functions – unintended effects on order/stability
Dysfunctions – disruptive consequences to society
manifest dysfunctions – anticipated disruptions to order/stability
latent dysfunctions – unintended disruptions to order/stability.

7. Use the following chart to summarize a functionalist analysis of community-wide celebrations.

	Function	Dysfunction
Manifest	markets public relations provides family activities unifies the community through shared experience	traffic jams garbage pile-ups no clean bathrooms closed streets
Latent	break down barriers across neighborhoods walk through park: up close observation transportation system	city workers negotiate contracts miss work due to a hangover.

8. What is the *maquiladora* program? Why was it established?

- encourage US and other foreign investers to locate plants along the border - employment opportunities
- US corporations get access to low-wage labour pool
- US workers no longer willing to fill low wage jobs

9. What are *colonias*?

Substandard settlements along the US/Mexican borders.

10. List one example of a manifest function, manifest dysfunction, latent function, and latent dysfunction associated with *maquiladoras*.

manifest - mexicans get jobs, US gets low wage labour
latent - political/social ties strengthen
manifest dys - job displacement, less skilled = no jobs
latent dys - large settlements w/no sanitation, water ect.

11. Summarize the strengths and weaknesses of the functionalist perspective.

12. Distinguish between the bourgeoisie and the proletariat. Which of the two is the exploiting class? Why? How is exploitation justified?

- proletariat - sell labour, own nothing
- bourgeoisie - own means of production, purchase labor
- Facade of legitimacy - explanation of members in the dominant groups to justify actions

ex: workers are free to leave

13. What are the major shortcomings of the conflict perspective?

- overstates the tensions/divisions and understates stabilities.
- neglect situations in which consumers, citizen groups, or workers use economic incentives to modify/control way capitalist persue profit.
- ignores contributions of industrialization in improving peoples standard of living.

14. How would a conflict theorist explain the purpose of the *maquiladora* program?

- maquilas ~~persue~~ represent pursuit of profit.
- means of production owned by capitalists in the US
- mexicans sell their labor, cannot demand higher wages or they will get replaced, gain very little.
- focus on exploitive conditions.

15. What central concepts and questions guide the symbolic interactionist perspective? What are the major shortcomings of this perspective?

- focus on how people make sense of the world, experience and define on how they influence/are influenced by others.
- concerned with how the self develops, how people attach meanings to actions and how they evolve.

16. How can the three theoretical perspectives help us to avoid hasty and oversimplistic reactions to headline news, sound bites, and other "facts" we hear?

- By using the scientific method - collection of data gained through observation and verification

17. Define research methods.

1. choose a topic 2. review the literature 3. identify core concepts 4. form hypothesis, collect data 5. analyze the data 6. Draw conclusions.

* objectivity, publish, cite *

18. What assumptions underlie the scientific method? Under what circumstances do research findings endure? Contrast the ideal of the research process with reality.

If a finding can be duplicated and repeated it is taken seriously

19. Why is it important for researchers to explain their reasons for choosing to investigate a particular topic?

- it clarifies the purpose/significance of the project
- what are the motivations

20. Why should researchers review the literature before beginning to investigate a topic?

- avoid repeating earlier work
- generates insight

21. What are concepts and how do they relate to the research process?

- thinking/communication tools used to give/receive complex info efficiently, frame/explain observations
- sociologists use their discipline concepts to give focus to what they choose to study.

22. What kinds of "populations" do sociologists study? Give examples.

traces - materials/physical evidence that yeilds info about human activity
documents - written/printed material used in research
territories - settings w/ borders - set aside for activities
households - related/unrelated who share same dwelling

23. Why do sociologists study random samples? Why are random samples difficult to secure? Under what conditions are nonrandom samples acceptable?

- every case has equal chance of being selected, non-biased
- nonrandom = have special characteristics

24. Give a brief description of each method of data collection.

Self-Administered Questionnaires - set of questions given to respondants who read instructions and fill in answers themselves. (exams, multiple choice)

Structured Interviews
face to face, telephone or internet between interviewer and respondant - ask q's and record answers. wording isn't changed

Unstructured Interviews
- question/answer sequence is spontaneous open-ended and flexible

Participant Observation
researchers interact directly with study participants

Secondary Sources
data collected by other researchers for some other pupose

25. What is the Hawthorne effect?

research subjects alter their behaviour when they learn they are being observed

26. What is an hypothesis? Give an example of an hypothesis. Identify the independent and dependent variables.

- Hypothesis: trial explanation, prediction,
ex this pill will cause this sort of reaction

independent - explains/predicts dependent (is m
dependent - to be explained or predicted.

27. What is an operational definition? Give an example.

- clear, precise definitions and instructions about how to observe and measure concepts and variables.

28. Explain "measurement is a prerequisite for good management" and "what gets measured gets managed".

29. Distinguish between reliability and validity.

- reliability - extend to which operational definition gives consistent results - repeatability
- validity - degree to which operational definition measures what it claims to measure

30. What is generalizability? Under what conditions can findings be considered generalizable?

- extent to which findings can be applied to the larger population from which it was drawn.
- sample is randomly selected, if the response rate for every question was high.

31. Distinguish among mean, mode, and median.

- mean: sum of scores divided by # of participants
- mode: the value that occurs most often
- median: the middle number 50% above, 50% below

32. What three conditions must be met before a researcher can claim that an independent variable contributes significantly to explaining a dependent variable?

- independent variable must precede dependent in time
- two variables must be correlated
- establishing a correlation is necessary but not in itself sufficient to prove causation.

Key Concepts

Theory	pg. 34
Sociological theory	pg. 34
Functionalist perspective	pg. 34
Function	pg. 34
Manifest functions	pg. 36
Latent functions	pg. 36
Dysfunction	pg. 37
Manifest dysfunctions	pg. 37
Latent dysfunctions	pg. 37

Concept Application

Consider the concepts listed below. Match one or more of the concepts with each scenario. Explain your choices.

√ Function
√ Dysfunction
√ Symbol
√ Territories
√ Participant observation
√ Spurious correlation
√ Proletariat

Scenario 1 "The influx of Korean-owned firms conferred obvious economic benefits on Los Angeles. (1) Korean firms tended to service low income, nonwhite neighborhoods generally ignored and underserved by big corporations....(2) The Korean influx restored the [deteriorating and underutilized] neighborhoods in which Koreans settled....(3) Their residential and commercial interests compelled Koreans to combat street crime....(4)Koreans valued public education and improved it. Indeed, many Korean families had emigrated to the United States because of this country's superior educational opportunities" (Light and Bonacich 1988:6-7).

Scenario 2 "Dr. Louise Keating became 'Trash Czar' for a few days. Dr. Keating, director of Red Cross Blood Services in Cleveland, found her center almost engulfed by mounds of debris—dressings, needles, plastic tubes—most of it the usual detritus of any organization, but some of it splashed with the blood of donors. Her center was not generating any more trash than usual. But suddenly no one was willing to cart it away. AIDS could be transmitted through blood, we had now learned. Last year's innocuous garbage had become this year's plague vector. Or so it seemed to Cleveland's carters. And the refuse piles grew.

Dr. Keating did solve her problem. Now, all waste that has any blood on it is sterilized in an autoclave until nothing, not even a virus, survives. But AIDS has created many other problems in the nation's blood supply: for those, like Dr. Keating and her colleagues, who must find donors and ensure that the blood obtained is safe; for those who give blood; and for those who receive it" (Murray 1990:205).

Scenario 3 Some Americans venturing into Mexico probably hear the word [gringo] and wonder if somebody is picking a fight. The answer seems to depend on who says it and how. "It's all in the tone; usually the eyes will tell you something as well," said Tony Garza, the U.S. ambassador to Mexcio, who grew up in Brownsville. "It can mean everything from "I am going to try and kick your butt," to "friend, let's have a drink," Garza added. "Let's jus say it is very situation-specific." When gringo is used in Mexico, it tends to be applied to anyone born in the United States, regardless of race or background (Schiller 2004).

Scenario 4 "I gained access to the enterprise through a friend who was a manager in a local bank from which the enterprise borrowed commercial loans. Management and workers in both factories knew I was a graduate student writing a dissertation. I was a full-time assembly worker in the Hong Kong plant, visited workers' homes, and participated in their weekend activities. In Shenzhen, I observed and talked with workers and managers on the shop floor and the office, but management allowed me to work on the line only occasionally. I lived in factory dormitories together with other Hong Kong managerial staff, but I visited and interviewed workers in workers' dormitories. I also participated in both workers' and managers' gatherings after work" (Lee 1995:380).

Scenario 5 For the class, the suburban mall became the microsocial setting for investigating macrotheoretical issues. Students examined specific features of their selected malls, such as the surrounding physical environment (entrance, parking, sidewalks), financial condition (unoccupied spaces, needed repairs, open-air merchants), design of interior space (escalators, lighting, plants), types of stores (prestige anchors, discounters, specialties), clientele (social class, gender, race, ethnicity), nationality as well as race/ethnicity and gender of merchants (especially subcontractors within stores) and employees, pricing structure (including types of credit cards accepted or interest-free purchase options), mall names and distinctive linguistic terms, treatment of shoppers by employees, safety/security issues, and the presence of "mall zombies" as a crude indicator of the dehumanizing effects associated with "irrationality of rationality" (Manning, Price, and Rich 1997:18).

Scenario 6 There is a positive correlation between ice cream sales and deaths due to drowning: the more ice cream sold, the more drownings, and vice versa. The third variable at work here is *season* or *temperature*. Most drowning deaths occur during the warm days of summer—and that's the peak period for ice cream sales. There is no direct link between ice cream and drowning, however. (Babbie 1995:70)

Scenario 7 In April, KenSa started production in a rented, temporary facility in San Pedro Sula while a new factory was built. It hired 150 workers to make Chrysler minivan door wire harnesses at the Honduran minimum wage of about 55 cents an hour. It turned hundreds of people away...But the jobs provide no economic miracle. Factory work barely provides enough to live...Most workers at foreign-owned factories in Honduras make $4.44 a day. That's less than the $5 a day Henry Fod paid his Highland Park workers 90 years ago, in 1914. Ford's pay (the equivalent of $11 an hour today) more than doubled the minimum wage at the time and helped give birth to America's blue-collar middle class...U.S. companies have no such incentive in countries such as Honduras. Products are built for export back to America. Raising worker salaries in San Pedro Sula won't sell even one more SUV in Detroit (French 2004).

Applied Research

Find a research article in a sociological journal such as *Sociological Focus, Journal of Comparative Sociology*, or *American Sociological Review*. Give a brief description of the following elements:

Topic
Major concepts or theoretical perspectives
Independent and dependent variables (if applicable)
Unit of analysis
Methods of data collection
Operational definition (if applicable)
Findings
Conclusion

Practice Test: Multiple-Choice Questions

1. Special attention is given to Mexico in the theory and research chapter because
   a. Mexico is an economic liability to U.S. taxpayers.
   b. Mexico is a major drug supplier.
   c. most Americans know little about the extent to which the U.S. depends on low-wage Mexican labor.
   d. people in the United States suffer from the NAFTA agreement.

2. The United States shares a ____________-mile long border with Mexico.
   a. 200
   b. 800
   c. 2,000
   d. 5,000

3. The question "Why does a particular arrangement exist?" is of greatest interest to a(n)
   a. functionalist.
   b. conflict theorist.
   c. symbolic interactionist.
   d. action theorist.

4. Sociologist Herbert Gans maintained that the poor serve many functions for society. Which one of the following is <u>not</u> one of those functions?
   a. Most give up looking for work and thus are not counted as unemployed.
   b. The poor do the most undesirable jobs in society at a low wage.
   c. The poor do time-consuming work for affluent persons.
   d. Certain middle-class occupations exist to serve the poor.

5. Community-wide celebrations can break down barriers across neighborhoods. The effect can be classified as a
   a. manifest function.
   b. latent function.
   c. manifest dysfunction.
   d. latent dysfunction.

6. A functionalist would ask which one of the following questions about *maquiladoras*?
   a. Why do *maquiladoras* exist, and what consequences do they have for the United States and Mexico?
   b. Who benefits from the *maquiladora* arrangement and at whose expense?
   c. Does everyone in the United States and Mexico see *maquiladoras* in the same way?
   d. Why are there no *maquiladoras* in the United States?

7. One purpose of the *Bracero* Program was to
   a. replace American workers who were serving in the Vietnam War.
   b. provide laborers for agricultural and other low-wage jobs to the United States.
   c. discourage Mexican immigration to the United States.
   d. discourage U.S. investment in Mexico.

8. An unexpected outcome of the *maquiladora* program is the rapid and unregulated population growth along the border. This outcome is an example of a
   a. manifest function.
   b. manifest dysfunction.
   c. latent function.
   d. latent dysfunction.

9. Marx wrote that the worker is an "appendage of the machine, and it is only the most simple, most monotonous, and most easily acquired knack that is required of him." Marx was describing the
   a. proletariat.
   b. the facade of legitimacy.
   c. bourgeoisie.
   d. dysfunctions.

10. The question "Who benefits from a particular pattern or social arrangement?" is of most interest to a(n)
    a. functionalist
    b. conflict theorist.
    c. symbolic interactionist.
    d. action theorist.

11. Which one of the following key words is associated with the conflict perspective?
    a. symbol
    b. façade of legitimacy
    c. order
    d. stability

12. In her study of Mexico's *maquiladoras*, Patricia Wilson describes three different environments and notes that
    a. workers have no opportunity to shape the production process.
    b. not all assembly plants organize the production process to exploit the worker.
    c. the work environment is very similar from one *maquila* to the next.
    d. *maquila* workers enjoy a better work environment than their U.S. counterparts.

13. Symbolic interactionists believe that during interaction, the parties involved
    a. respond directly to the surroundings and to each other's actions.
    b. must share a symbol system if they are to communicate.
    c. usually communicate effectively even if they do not share the same symbol system.
    d. do not have to share a symbol system.

For the following statements, identify which theoretical perspective is associated with each.

a. Functionalist
b. Conflict
c. Symbolic interaction

14. Mexico's GNP would drop by $3.1 billion if the *maquiladora* industry ceased to exist.

15. Lack of Spanish-speaking skills and lack of experience with foreign cultures lay at the heart of expatriate-Mexican local clashes.

16. It wasn't until the peso devaluation of 1998 that the *maquiladora* industry exploded into a great industry.

17. Sociological research is guided by
    a. techniques and strategies unique to the discipline.
    b. a passion to change society.
    c. emotion and personal interest.
    d. the scientific method.

18. In theory, the first step in undertaking a sociological research project is
    a. consulting existing research.
    b. collecting data.
    c. choosing a topic for investigation.
    d. analyzing the data.

19. __________ are the unit of analysis for a research project that collects garbage from landfills and selected neighborhood garbage cans.
    a. Traces
    b. Documents
    c. Territories
    d. Households

20. The U.S. census, which is mailed out to every household every 10 years, is an example of a(n)
    a. experiment.
    b. observation.
    c. self-administered questionnaire.
    d. secondary research.

21. _________________ is especially useful for studying behavior as it occurs.
    a. Self-administered questionnaires
    b. Secondary data analysis
    c. Interviews
    d. Observation

22. A dependent variable is
    a. the variable of cause.
    b. a trial idea.
    c. the variable to be explained.
    d. the core concept.

23. The *independent* variable in the hypothesis "the closer a *maquiladora* is located to the U.S.-Mexican border, the higher the percentage of U.S. patriots working at the plant," is
    a. *maquilas* distance to the U.S.-Mexican border.
    b. percentage of U.S. expatriates working at the plant.
    c. U.S.-Mexican border.
    d. number of workers at the plant.

24. The question "Is this operational definition really measuring what it claims to measure?" addresses concerns surrounding
    a. sampling.
    b. validity.
    c. reliability.
    d. correlations.

25. ________________________ is one of the best predictors of a company's economic performance.
    a. Employee retention
    b. Profits
    c. Sales
    d. Revenue generated per employee

26. For each of the 254 counties in the state of Texas, Erin finds the percentage of people in each county classified as Hispanic. She finds that on average, each county has an Hispanic population of 23.6 percent. Erin has found the
    a. mode.
    b. median.
    c. mean.
    d. standard deviation.

27. The correlation between the variables "percentage of population living at poverty level or below" and "percentage of population classified as Hispanic" is +0.60. This means the ________ the percentage of the population living at poverty or above the _________ the percentage of population classified as Hispanic.
    a. greater; greater
    b. greater; lower
    c. lower; greater
    d. fewer; higher

28. Which one of the following statements best describes how the three perspectives should be viewed?
    a. A single perspective can give us a complete picture of a process or an event.
    b. Most sociologists maintain that one perspective only should be adopted when analyzing an issue.
    c. The three perspectives should be viewed as opposing viewpoints.
    d. We can gain greater understanding of a process or an event if we examine it from the point of view of more than one perspective.

29. A functionalist reads the headline "U.S. strengthens patrols along Mexican border." He or she would begin to think about
    a. how the presence of border guards contributes to order and stability in both societies.
    b. interactions between border guards and Mexican illegals.
    c. who benefits from the existence of border guards, and at whose expense.
    d. identifying policies to stop the flow.

Practice Test: True/False Questions

T F 1. The United States is Mexico's largest trading partner.

T F 2. A functionalist would argue that sports teams have no real purpose in society.

T F 3. Workers in rural U.S. counties are particularly vulnerable to job displacement associated with outsourcing.

T F 4. Conflict theorists draw their inspiration from George Herbert Mead.

T F 5. A down-to-earth research approach means that sociologists immerse themselves in the social world chosen for study.

T F 6. Research findings endure if they can withstand continued reexamination and duplication by scientists.

T F 7. Researchers can manipulate data if the deception supports well-intentioned personal, economic, and political agendas.

T F 8. Researchers most often study entire populations (as opposed to samples).

T F 9. Approximately one-half of Fortune 500 companies report that they have moved service jobs outside the United States.

T F 10. A correlation of -1.0 suggests there is no relationship between two variables.

Internet Resources Related to Theoretical Perspectives and Research Methods

- **Dead Sociologists Society**
  http://www2.pfeiffer.edu/~lridener/DSS/DEADSOC.HTML
  Larry Ridener, Department of Sociology at Baylor, gives an overview of the lives and works of Comte, Martineau, Marx, Spencer, Durkheim, Simmel, Weber, Veblen, Addams, Cooley, Mead, Thomas, Du Bois, Pareto, Sorokin and Park.

- **Journal of Statistics Education**
  http://www.amstat.org/publications/jse/
  The *Journal of Statistics Education* "provides interesting and useful information, ideas, software, and data sets" that help explain various techniques of data analysis.

Internet Resources Related to Mexico

- **US/Mexico Border Counties Coalition**
  http://www.bordercounties.org/
  The United States/Mexico Border Counties Coalition is a "nonpartisan, consensus-based policy and technical forum founded to address challenges facing county governments located on the United States/Mexico Border. The elected officials from the twenty-four county governments located on the United States/Mexico Border established the Coalition. These county governments face unique challenges in serving their residents."

Mexico: Background Notes

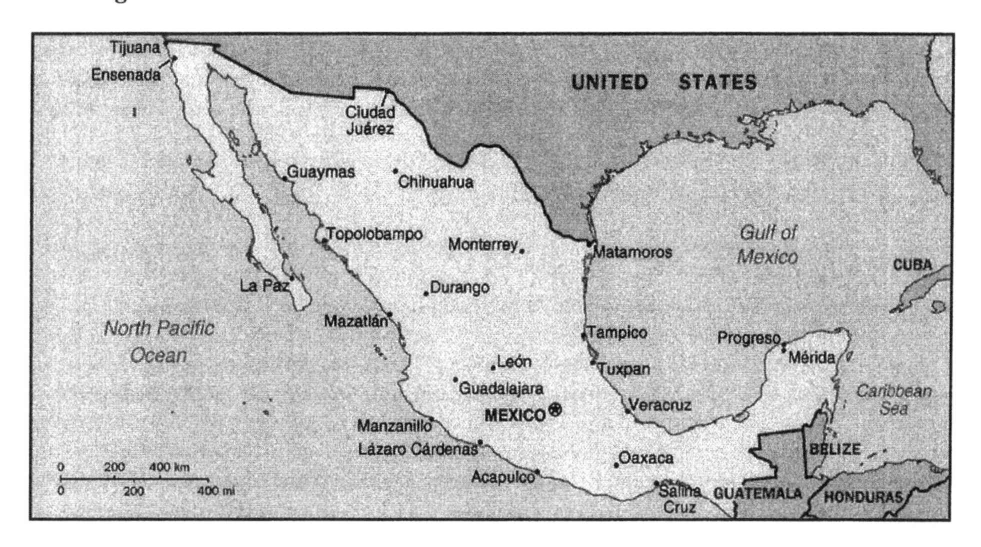

PEOPLE

Nationality: Noun and adjective—Mexican(s).
Population (2004 estimate): 105 million.
Annual growth rate (net) 2004: 1.2%.
Ethnic groups: Indian-Spanish (mestizo) 60%, Indian 30%, Caucasian 9%, other 1%.
Religions: Roman Catholic 89%, Protestant 6%, other 5%.
Language: Spanish.
Education: Years compulsory--12 (note: preschool education was made mandatory in Dec. 2001). Literacy--89.4%.

Health (2004 est.): Infant mortality rate—21.69/1000. Life expectancy—male 72.18 years; female 77.83 years.
Work force (2000, 39.81 million): Agriculture, forestry, hunting, fishing--21.0%; services--32.2%; commerce--16.9%; manufacturing--18.7%; construction--5.6%; transportation and communication--4.5%; mining and quarrying--1.0%.

Mexico is the most populous Spanish-speaking country in the world and the second most-populous country in Latin America after Portuguese-speaking Brazil. About 70% of the people live in urban areas. Many Mexicans emigrate from rural areas that lack job opportunities—such as the underdeveloped southern states and the crowded central plateau—to the industrialized urban centers and the developing areas along the U.S.-Mexico border. According to some estimates, the population of the area around Mexico City is about 18 million, which would make it the largest concentration of population in the Western Hemisphere. Cities bordering on the United States—such as Tijuana and Ciudad Juarez—and cities in the interior—such as Guadalajara, Monterrey, and Puebla—have undergone sharp rises in population in recent years.

Education is one of the Government of Mexico's highest priorities. The education budget has increased significantly in recent years; funding in real terms for education has increased by almost 25% over the last decade. Education in Mexico also is being decentralized from federal to state authority in order to improve accountability. Although educational levels in Mexico have improved substantially in recent decades, the country still faces daunting problems.

Education is mandatory from ages 6 through 18. In addition, the Mexican Congress voted in December of 2001 to make one year of preschool mandatory, which went into effect in 2004. The increase in school enrollments during the past two decades has been dramatic. By 1999, 94% of the population between the ages of 6 and 14 were enrolled in school. Primary, including preschool, enrollment totaled 17.2 million in 2000. Enrollment at the secondary public school level rose from 1.4 million in 1972 to 5.4 million in 2000. A rapid rise also occurred in higher education. Between 1959-2000 college enrollments rose from 62,000 to more than 2.0 million.

HISTORY

Highly developed cultures, including those of the Olmecs, Mayas, Toltecs, and Aztecs existed long before the Spanish conquest. Hernando Cortes conquered Mexico during the period 1519-21 and founded a Spanish colony that lasted nearly 300 years. Independence from Spain was proclaimed by Father Miguel Hidalgo on September 16, 1810; this launched a war for independence. An 1821 treaty recognized Mexican independence from Spain and called for a constitutional monarchy. The planned monarchy failed; a republic was proclaimed in December 1822 and established in 1824.

Prominent figures in Mexico's war for independence were Father Jose Maria Morelos; Gen. Augustin de Iturbide, who defeated the Spaniards and ruled as Mexican emperor from 1822-23; and Gen. Antonio Lopez de Santa Ana, who went on to control Mexican politics from 1833 to 1855. Santa Ana was Mexico's leader during the conflict with Texas, which declared itself independent from Mexico in 1836, and during Mexico's war with the United States (1846-48). The presidential terms of Benito Juarez (1858-71) were interrupted by the Habsburg monarchy's rule of Mexico (1864-67). Archduke Maximilian of Austria, whom Napoleon III of France established as Emperor of Mexico, was deposed by Juarez and executed in 1867. Gen. Porfirio Diaz was president during most of the period between 1877 and 1911.

Mexico's severe social and economic problems erupted in a revolution that lasted from 1910-20 and gave rise to the 1917 constitution. Prominent leaders in this period—some of whom were rivals for power—were Francisco I. Madero, Venustiano Carranza, Pancho Villa, Alvaro Obregon, Victoriano Huerta, and Emiliano Zapata. The Institutional Revolutionary Party (PRI), formed in 1929 under a different name, emerged as a coalition of interests after the chaos of the revolution as a vehicle for keeping political competition in peaceful channels. For 71 years, Mexico's national government had been controlled by the PRI, which had won every presidential race and most gubernatorial races until the July 2000 presidential election of Vicente Fox Quesada of the National Action Party (PAN).

ECONOMY

Nominal GDP (2003 est.): $615 billion. (7.4 trillion pesos, 2004 Q2)
Per capita GDP (2003 est.): $5,945.
Annual real GDP growth 2003: (1.3%); 2002 (0.9%); 2001 (-0.3%); 2000 (6.6%); 1999 (3.7%).
Avg. real GDP growth (1999-2003): 2.1%.
Inflation rate: 2003 (4.0%); 2002 (5.0%); 2001 (6.4%); 2000 (9.5%); 1999 (16.6%).
Natural resources: Petroleum, silver, copper, gold, lead, zinc, natural gas, timber.

Agriculture (5.8% of GDP): Products—corn, beans, oilseeds, feed grains, fruit, cotton, coffee, sugarcane, winter vegetables.
Industry (26.0% of GDP): Types—manufacturing, petroleum, and mining.
Services (68.3% of GDP): Types—commerce and tourism (18%), financial services (11%), and transportation and communications (10%).
Trade (Goods): Exports (2003)--$164.9 billion. Imports (2003)--$170.5 billion. Exports to U.S. (2003)--$144.5 billion. Imports from U.S. (2003)--$105.4 billion. Major markets (2003)--U.S. (in 2003 destination for 88% of Mexico's exports; in 2003 source for 62% of Mexico's imports), EU, Japan, Canada, China, other significant trade partners.

Mexico is highly dependent on exports to the U.S., which account for almost a quarter of the country's GDP. The result is that the Mexican economy is strongly linked to the U.S. business cycle. As the U.S. economy has emerged from its downturn in 2001, so has the Mexican economy, growing at a 3.8% rate in the first half of 2004.

Mexican trade policy is among the most open in the world, with Free Trade Agreements with the U.S., Canada, the EU, and many other countries. Since the 1994 devaluation of the peso Mexican governments have improved the country's macroeconomic fundamentals. Inflation and public sector deficits are both under control. As of September 2004, Moody's, Standard & Poors, and Fitch Ratings have all issued investment-grade ratings for Mexico's sovereign debt.

TRADE

Mexico is one of the world's most trade dependent countries, and it is particularly dependent on trade with the U.S, which buys approximately 88% of its exports. Top U.S. exports to Mexico include electronic equipment, motor vehicle parts, and chemicals. Top Mexican exports to the U.S. include petroleum, cars, and electronic equipment. There is considerable intra-company trade.

Mexico is an active and constructive participant in World Trade Organization (WTO) matters, including in the launching of the Doha trade route. Mexico hosted the WTO Ministerial Meeting in Cancun in September 2003. The Mexican Government and many businesses support a Free Trade Area of the Americas.

Trade disputes between the U.S. and Mexico are generally settled in WTO or North American Free Trade Agreement (NAFTA) panels or through negotiations between the two countries. The most significant areas of friction involve agricultural products including sugar, high fructose corn syrup, apples, and rice.

MANUFACTURING AND FOREIGN INVESTMENT

Manufacturing accounts for about 20.3% of GDP and grew by 9.4% in 2000. Manufacturing probably fell or was stagnant in 2001 because exports to the U.S. probably fell. Construction grew by almost 7% in 2000 but was probably stagnant in 2001.

According to Mexico's Ministry of Economy, Foreign Direct Investment (FDI) in Mexico for 2003 was $10.38 billion, down 29% from the year before. The U.S. was once again the largest foreign investor in Mexico, with $5.75 billion in investments, or 55% of total FDI. The most recent numbers released by Mexico show FDI for January thru June 2004 at $9.57 billion. Although the amount is nearly equal to all of 2003, the total is inflated by an investment of over $4.0 billion by the Spanish bank BBVA.

OIL AND GAS

In 2003 Mexico was the world's fifth-largest oil producer, its 9th- largest oil exporter, and the third-largest supplier of oil to the United States. Oil and gas revenues provide about one-third of all Mexican Government revenue.

Mexico's state-owned oil company, Pemex, holds a constitutionally established monopoly for the exploration, production, transportation, and marketing of the nation's oil. Since 1995, private investment in natural gas transportation, distribution, and storage has been permitted, but Pemex remains in sole control of natural gas exploration and production. Despite substantial reserves, Mexico is a net natural gas importer.

U.S.-MEXICAN RELATIONS

U.S. relations with Mexico are as important and complex as with any country in the world. A stable, democratic, and economically prosperous Mexico is fundamental to U.S. interests. U.S. relations with Mexico have a direct impact on the lives and livelihoods of millions of Americans—whether the issue is trade and economic reform, homeland security, drug control, migration, or the promotion of democracy. The U.S. and Mexico are partners in NAFTA, and enjoy a rapidly developing trade relationship.

The scope of U.S.-Mexican relations goes far beyond diplomatic and official contacts; it entails extensive commercial, cultural, and educational ties, as demonstrated by the annual figure of nearly a million legal border crossings a day. In addition, more than a half-million American citizens live in Mexico. More than 2,600 U.S. companies have operations there, and the U.S. accounts for 55% of all foreign direct investment in Mexico. Along the 2,000-mile shared border, state and local governments interact closely.

There is frequent contact at the highest levels. The Presidents' meetings have included the Asia-Pacific Economic Cooperation Summit in Bangkok in October 2003, President Bush's visits to Monterrey in January 2004 (Summit of the Americas) and March 2002; his April 2001 visit to Guanajuato; and President Fox's state visit to the U.S. in September 2001 and his meeting with the President in Crawford, Texas in March 2004.

Since 1981, the management of the broad array of U.S.-Mexico issues has been formalized in the U.S.-Mexico Binational Commission, composed of numerous U.S. cabinet members and their Mexican counterparts. The commission holds annual plenary meetings, and many subgroups meet during the course of the year to discuss border security and counter terrorism, trade and investment opportunities, financial cooperation, consular issues and migration, legal affairs and anti-narcotics cooperation, education, energy, border affairs, environment and natural resources, labor, agriculture, health, housing and urban development, transportation, and science and technology.

A strong partnership with Mexico is critical to combating terrorism and controlling the flow of illicit drugs into the United States. Cooperation on counter-narcotics and Mexico's own initiatives in fighting drug trafficking have been unprecedented. The U.S. will continue working with Mexico to help ensure that Mexico's cooperation and anti-drug efforts grow even stronger. The U.S. and Mexico continue to cooperate on narcotics interdiction, demand reduction, and eradication.

Source: U.S. Department of State (2004)

Chapter References

Haub, Carl and Machiko Yanagishita. 1994. *1994 World Population Data Sheet*. Washington, DC: Population Reference Bureau.

Herzog, Sergio. 2003. "Does the Ethnicity of Offenders in Crime Scenarios Affect Public Perceptions of Crime Seriousness? A Randomized Survey Experiment in Israel." *Social Forces* 82:2(757).

Light, Ivan and Edna Bonacich. 1988. *Immigrant Entrepreneurs: Koreans in Los Angeles 1965-1982*. Los Angeles: University of California Press.

Murray, Thomas H. 1990. "The Poisoned Gift: AIDS and Blood." *The Milbank Quarterly* 68(2):205-25.

Samora, Julian, Lyle Saunders, and Richard F. Larson. 1965. "Medical Vocabulary Knowledge Among Hospital Patients." Pp. 278-91 in *Social Interaction and Patient Care* edited by J. K. Skipper, Jr. and R. C. Leonard. Philadelphia: Lippincott.

U.S. Department of State. 2004. "Background Notes: Mexico" www.state.gov.

Answers

Concept Application

1. Function pg. 34
2. Dysfunction pg. 37
3. Symbol pg. 49
4. Participant observation pg. 58
5. Territories pg. 54
6. Spurious correlation pg. 65
7. Proletariat pg. 45

Multiple-Choice

1. c pg. 32
2. c pg. 32
3. a pg. 34
4. a pg. 35
5. b pg. 37
6. a pg. 38
7. b pg. 38
8. d pg. 40
9. a pg. 45
10. b pg. 43
11. b pg. 45
12. b pg. 47
13. b pg. 49
14. a pg. 39
15. c pg. 51
16. b pg. 47
17. d pg. 52
18. c pg. 52
19. a pg. 54
20. c pg. 56
21. d pg. 57
22. c pg. 59
23. a pg. 59
24. b pg. 61
25. a pg. 60
26. c pg. 63
27. a pg. 65
28. d pg. 66
29. a pg. 67

True/False

1. T pg. 32
2. F pg. 34
3. T pg. 40
4. F pg. 44
5. T pg. 49
6. T pg. 52
7. F pg. 52
8. F pg. 55
9. T pg. 58
10. F pg. 64

Chapter 3

Culture
With Emphasis on North and South Korea

Study Questions

1. Why were the Koreas chosen to illustrate the concept of culture?

KoreanWar Have had a profound effect on relationship between US and Korea.
What it means to be korean

2. According to the *Cambridge International Dictionary of English*, how is the word *culture* typically used among English speakers?

~~custom~~ def: the way of life, general customs and beliefs of a particular group of people at a particular time.
- used in conjunction with specific places and people
- used to emphasize differences

3. What questions highlight the challenges connected with defining and studying culture? Use the cases of Wayne Berry and Andrea "Simone" Bowers to illustrate the challenges.

- is it possible to define something something so vast as way of life.
- who belongs to what group (korean or American?) ?
- do certain characterstics separate cultures (eating rice in morning)

?

4. Distinguish between nonmaterial and material culture.

Material: all physical objects that people have borrowed, discovered, or invented and to which they attach meaning.
Ex: trees, plants, minerals, cars, radios, computers ect.
Nonmaterial: Intangible human creations that we cannot identify directly through the senses
Ex: beliefs, values and norms.

5. When sociologists study material culture, what questions do they ask?

what are it's uses? what are the meanings assigned by the people who use it? How does it shape social relationships?

Ex: Radio - plays music - fills loneliness
Microwave - cuts down cooking time - less family dinners.

6. Distinguish between beliefs and values. Give example of each.

Beliefs: conceptions that people accept as true, concerning how the world operates and where the individual fits in relationship to others. Ex: Athletic talent is inherited.
Values: General, shared conceptions of what is good, right, appropriate, worthwhile, important in regard to conduct, appearance, states of being.
Ex: freedom, happiness, obedience, cleanliness.

7. What are norms? Distinguish between folkways and mores.

- written/unwritten rules that specify behaviors appropriate and inappropriate to a particular social situation.

Folkways: norms that apply to the mundane aspects or details of daily life. (when and what to eat)

Mores: norms that people define as critical to the wellbeing of a group. violation of mores result in severe forms of punishment.

8. How do geographical/historical forces influence culture?

- material/nonmaterial aspects of culture represents the solutions worked out to meet geographical/historical challenges.
  Ex: amount of available resources.
- conservation and consumption orientated values.

9. Explain "Regardless of their physical traits babies are destined to learn the ways of the culture into which they are born and raised."

- our genes endow us with our human and physical characteristics, but not our cultural characteristics.
- People behave as they do because these behaviours seem natural, we lose sight that culture is learned.

10. How is language a thinking tool?

- As we learn words and meanings in language we learn about culture.
- interpret experiences, establish/maintain relationships and convey information.
- conveys messages beyond the meaning of the words (connotations)

11. How are people products of cultural experiences yet not cultural replicas of one another?

- People have a choice of cultural options
- Exposed to different "versions" of culture - pass on to children ect.
- can reject, manipulate, create culture - not passive agents

12. What are some of the ways in which culture stimulates and satisfies appetite?

Different foods available make different food items "edible"
Ex: Rice - South Korea, corn - USA. Dogmeat vs. Beef.
- who cooks, how food is served, cooked ect.

13. What makes an emotion social? What is the connection between feeling rules and social emotions?

Def: internal bodily sensations that we experience in relationships with other people.

Ex: Empathy, grief, love, guilt, jelousy, embarrassment

Feeling rules: Norms that specify appropriate ways to express internal body sensations.

14. What is the relationship between material and nonmaterial cultures? Give an example.

-People ~~borrow~~ borrow both non- and material cultures from other societies. Ex Japanese McDonalds.

15. What is diffusion? Give two examples of the diffusion process. Why is diffusion a selective process?

Diffusion = process by which an idea, invention or some other cultural item is borrowed from a foreign source.
Ex: Face to face, or via Internet.
- people are choosy about which items they adopt, can modify it to fit that cultures norms/values.

16. Give two examples of opportunities for cultural diffusion between Americans and South Koreans.

17. What is culture shock? How is it related to ethnocentrism?

The strain that people from one culture experience when they must reorient themselves to the ways of a new culture.
- people accept that their cultural is natural

18. What are the various types of ethnocentrism? Give examples of each. (Don't forget reverse ethnocentrism.)

Cultural genocide: people of one society define another as intolerable and they seek to destroy it.
Reverse ethnocentrism: home culture is regarded as inferior to a foreign culture

19. Explain reentry shock.

Culture shock in reverse, it is experienced upon returning home after living in another culture

20. What viewpoint should one take when studying other cultures?

Cultural relativism

21. Is cultural relativism equivalent to moral relativism? Explain.

- foreign culture should not be judged by homec culture
- behaviour/way of thinking must be examined in it's cultural context

22. What are subcultures? When are subcultures institutionally complete?

- Groups that share in some parts of the dominant culture but have their own distinctive values, norms, language and/or material culture
- members do not interact with anyone outside their subculture - their subculture satisfies all their needs.

23. Does everyone who lives in a particular country share the same culture? What concepts do sociologists use to distinguish groups who depart from the so-called mainstream culture?

24. How do the facts of cultural diffusion and individual autonomy expand our conception of culture?

Key Concepts

Material culture	pg. 75
Nonmaterial culture	pg. 76
Beliefs	pg. 76
Values	pg. 76
Norms	pg. 77
Folkways	pg. 77
Mores	pg. 80
Feeling rules/ Social emotions	pg. 88
Language	
Denotation	pg. 84
Connotation	pg. 84
Idiom	pg. 84

Diffusion	pg. 90
Ethnocentrism	pg. 96
Cultural genocide	pg. 96
Reverse ethnocentrism	pg. 98
Culture shock	pg. 92
Re-entry shock	pg. 95
Cultural relativism	pg. 99
Subcultures	pg. 100
Institutionally complete	pg. 100

Concept Application

Consider the concepts listed below. Match one or more of the concepts with each scenario. Explain your choices.

- ✓ Diffusion
- ✓ Feeling rules
- ✓ Norms
- ✓ Reentry shock
- ✓ Reverse ethnocentrism
- ✓ Subcultures

Scenario 1 "Overseas, the home country environment becomes irrationally glorified. All difficulties and problems are forgotten and only the good things back home are remembered. Upon returning to the United States, people may be surprised to find that they not only miss their host country and its people, culture and customs, but also the people with whom they shared the experience. They realize how well they actually got along under a different set of living conditions and how much happened and changed back home in their absence. As one woman said to me, 'Three years is a long time to be immersed in another way of life and I felt numb and kind of left out or not 'in' on things happening in the United States. It was a very unhappy time for me because I had expected to be ecstatic to get home'" (Koehler 1986:90).

Scenario 2 "Yesterday, my 4-year-old stopped crying. He fell of his bike, held his breath and gritted his teeth. 'I'm not gonna cry, Mom,' he said. 'I'm really not.'

Where did this pint-size stoicism come from? Batman videos? Preschool name-callers? Maybe the neighbors who tell their kid, 'Crying will get you nowhere.' You hear it everywhere. You'd better not pout, you'd better not cry. Big boys don't cry. Grin and bear it, hide it, stifle it, but whatever you do, don't cry, *please*, don't cry. I'll give you a cookie if you stop" (Hogan 1994:E1).

Scenario 3 "The Hurricane Girls signal a new era of bikers. Far from the stereotypical motorcycle clubs filled with tattooed tough guys, they are all female, primarily African American and—in perhaps the greatest heresy to the old-school bikers—welcome bikes that aren't Harley-Davidsons.

The Bay Area has long been a hotbed of motorcycling and home to Oakland's East Bay Dragons, one of the oldest and most influential black clubs, as well as some of the most notorious chapters of the Hells Angels. But as those guys gray, a new brand of biker is emerging and clubs reflect that. There are gay clubs, clean-and-sober clubs, Buddhist clubs, Persian clubs and even Goth clubs. But many new clubs were founded by, and for, African Americans.

There are at least 40 black bike clubs locally, many of which surfaced in the past five years. The trend is catching on nationwide. Last year, the National Association of Black Bikers formed to unite black clubs across the country, regulate activities and help those starting out. The association, based in Maryland, includes more than 1,000 clubs representing 5,000 members." (Fullbright 2004)

Scenario 4 "The peanut is a South American original, a shrub whose flowers send tendrils into the ground where they grow into seed pods. It was domesticated about 4,000 years ago in the eastern foothills of the Andes, somewhere around the border between Bolivia and Argentina. And very thoroughly domesticated; it's one of the few plants in the world that is never found in the wild. By the time of Columbus, it had spread throughout Latin America and the Caribbean.

It was in Africa that the peanut found its heartiest welcome. The Portuguese brought it to West Africa in the early 1500s, and in 1564 the traveler Alvares de Almada reported it was already an established crop in Senegal and Gambia (which are still among the world's greatest peanut-exporting countries). Within 200 years it had spread on its own all the way across Africa to Angola, without Portuguese help" (Perry 1994:H9).

Scenario 5 "Japanese frequently bow to one another, for instance when greeting someone, as a gesture of respect and sincerity. The type of bow depends on the formality of the situation, the type of personal relationship (e.g., close or distant), and the differences in social status of the individuals involved. The bow might be no more than a simple nod of the head or, on more formal occasions, a deeper bow from the waist. The most formal bow involves kneeling, placing one's hands out in front on the floor, and lowering the head slowly so that it almost touches the floor. Bowing is not always required, however. Family members and close friends do not usually bow to each other, but a child might bow to his or her mother when apologizing for mischievous behavior" (Japan Information Center 1988:61).

Applied Research

One way to gain a firsthand and personal appreciation for the comparative approach is to talk with an international student attending your college. Once you find out what country the student is from, how he or she came to attend this college, and his or her academic interests, ask the following questions:

1. Before you came to the United States, what did you expect it to be like? How did it differ from your expectations?

2. What do you miss most? (or: What was the biggest adjustment that you had to make, once you arrived in the United States?)

3. What do you like most about life in the United States? Least?

4. Are there any words or expressions in your native language that you cannot find in English words? Can you give some examples? Are there any English words or expressions for which you can find no equivalent words in your language?

Practice Test: Multiple-Choice Questions

1. Currently there are approximately ______ U.S. military personnel stationed in North Korea.
   a. 10,000
   b. 40,000
   c. 83,000
   d. 100,000

2. Which one of the following descriptions does not apply to North Korea:
   a. communist-style government
   b. isolated
   c. centrally-planned economy
   d. republic

3. Which one of the following is not one of the challenges sociologists face in studying culture?
   a. describing culture
   b. determining who belongs to a group designated as a culture
   c. identifying the distinguishing characteristics that set one culture apart from another
   d. identifying the essential principles defining the broad nature of culture

4. In thinking about material culture, it is most important to consider
   a. the most obvious and practical uses.
   b. the inventor.
   c. the country of origin.
   d. the meanings assigned to it by the people who use it.

5. Signs that read "No Smoking," "Honk Horn to Open," and "Emergency Exit Only" specify
   a. values.
   b. norms.
   c. beliefs.
   d. mores.

6. In Korea, diners reach and stretch across one another and use their chopsticks to take food from serving bowls. These behaviors represent
   a. values.
   b. norms.
   c. beliefs.
   d. expressive symbols.

7. A sociologist seeking to explain why Koreans turn off their headlights while stopped at red traffic lights in the city would explore
   a. genetic attributes of Korean people.
   b. Korean norms.
   c. Korean values.
   d. geographic and historical factors.

8. In her essay "Group Study, Cheating, and Korean Student Experience" Kim Bo-Kyung observed that
   a. American universities require students to take a fixed course sequence.
   b. most American students join study groups.
   c. American students place a high value on "solitary effort".
   d. American teaching assistants were impressed with the Korean habits of study.

9 Part of the reason Koreans and Americans open refrigerators differently has to do with
   a. genetic differences between Americans and Koreans.
   b. the fact that Koreans are conservative by nature.
   c. the amount of natural resources in each country.
   d. the fact that Americans are incapable of conserving energy.

10. Korean-American youths who participate in cultural immersion programs that involve study in Korea often observe that "they never felt so American as when they are slurping noodles in Korea. Even their slurps have American accents." This example suggests that
    a. our genes endow us with our cultural characteristics.
    b. we can assume that someone comes from a particular culture if we know their physical characteristics.
    c. regardless of their physical traits people learn the ways of culture into which they are born and raised.
    d. humans are born with cultural characteristics.

11. "On the edge of my seat," and "To be in hot water'" are examples of
    a. denotations.
    b. folkways.
    c. idioms.
    d. cultural values.

12. In the United States singular possessive pronouns (e.g., "my") are used to refer to things we do not have exclusive control over. This reflects the American preoccupation with
    a. the group.
    b. the individual.
    c. competitiveness.
    d. resources.

13. One indicator of culture's influence on satisfying hunger is that
    a. only a portion of the potential food available in a society is defined as edible.
    b. people everywhere eat three meals a day.
    c. fast food appeals to people everywhere.
    d. if people are hungry enough, they will eat just about anything.

14. Which one of the following is <u>not</u> considered a social emotion?
    a. jealousy
    b. love
    c. drowsiness
    d. embarrassment

15. The U.S. Army publishes a list of "Must Know Items" about South Korea for American soldiers who are stationed there. One item says "Don't be surprised if you see two Korean women or men walking arm in arm. They are just good friends and there is nothing sexual implied." The Army is alerting soldiers to
    a. material culture.
    b. feeling rules.
    c. reverse ethnocentrism.
    d. idioms.

16. People of one society borrow ideas, materials, or inventions from another society
    a. if the items are exotic.
    b. if the people admire the country from which the items are borrowed.
    c. in selective ways.
    d. if the items can be acquired at a reasonable cost.

17. There are a number of striking differences between baseball as played in the U.S. and Japan. Which of the following is one of those differences?
    a. In Japan, controversial calls can take up to five hours to resolve.
    b. In Japan, there is a time limit on games, no matter the score.
    c. In Japan, there are five bases.
    d. In Japan, pitchers must get 10 batters out to end an inning.

18. Most people take an ethnocentric view toward foreign cultures; that is, they
    a. use their home culture as the standard for judging the worth of another culture.
    b. use a foreign culture as the standard to judge all others.
    c. judge a cultural practice in light of that society's values, environmental challenges, and history.
    d. seek to destroy the symbols of foreign cultures.

19. Under Japanese rule Korean students were taught by Japanese teachers, Korean names were changed to Japanese names, and practically everything Korean was abandoned. The Japanese were guilty of
    a. cultural relativity.
    b. institutional completeness.
    c. reverse ethnocentrism.
    d. cultural genocide.

20. Often there is a very clear association between __________ and institutional completeness.
    a. age
    b. ethnicity
    c. language
    d. income

True/False Questions

T F 1. Sociologists point out that people from the same culture are cultural replicas of one another.

T F 2. Much of Korean identity is intricately linked with the idea of being "not Japanese".

T F 3. The Statue of Liberty is a gift to the United States from Japan.

T F 4. Social psychologist Milton Rokeach identified only four values that people everywhere share to some degree.

T F 5. Jehovah Witnesses trace their beginnings to a small group of Bible students in Pittsburg, Pennsylvania.

T F 6. In Korea it is virtually impossible to carry on a conversation without referring to age.

T F 7. Jokes translate easily from one language to another.

T F 8. Travelers are more likely to prepare for the experience of culture shock than to prepare for the experience of reentry shock.

T F 9. Koreans breed poodles, collies, and German Shepherds not to be pets, but for food.

T F 10. The people of the two Koreas have almost no relationship with one another.

Internet Resources Related to Culture

- **ePALS Classroom Exchange**
  http://www.epalscorp.com/
  ePales is "internationally recognized as the leading provider of school-safe email and collaborative technology. Used in classrooms in **191 countries**, ePALS's multilingual network has made it possible for more than 4.6 million students and educators to employ the Internet as the ultimate communication and cross-cultural learning tool."

- **UNESCO-World Heritage**
  http://whc.unesco.org
  The UNESCO World Heritage Committee offers a list of **788** sites of "outstanding universal value" protected worldwide. UNESCO takes the position that "deterioration or disappearance of any item of cultural or natural heritage constitutes a harmful impoverishment of all the nations of the world.

Internet Resources Related to Korea

- **Korea Now**
  http://kn.koreaherald.co.kr
  Korea Now is a bi-weekly magazine that publishes articles related to Korean politics and policy, business and finance, and society and arts.

- **Korea Web Weekly**
  http://www.kimsoft.com/korea.htm
  Korea Web Weekly is "a non-partisan, non-profit news magazine dedicated to various issues related to Korea." Volunteers contribute news summaries, links and editorials. Volunteers are from a wide range of backgrounds including professors, Korean war veterans, expatriates, and so on. Some examples of interesting topics/links include "an American's View of U.S. Troops in South Korea."

Statistical Profile: Republic of South Korea

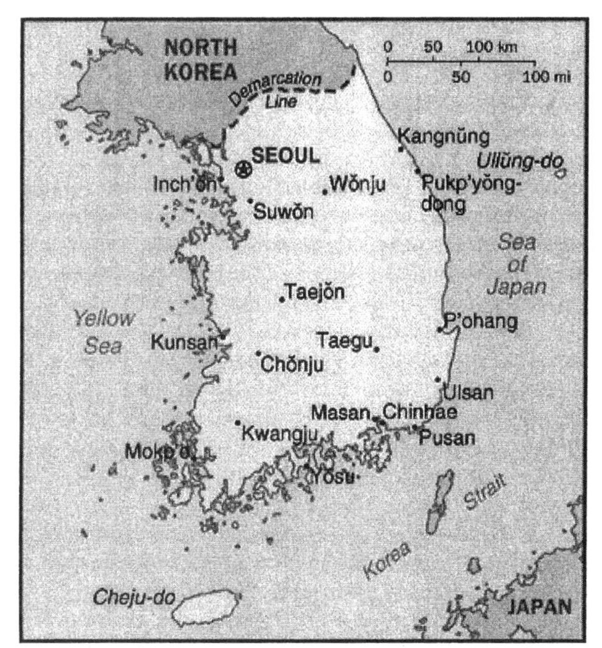

PEOPLE

Nationality: *Noun and adjective*--Korean(s).
Population (2000): 48.3 million.
Annual growth rate (2000): 0.93%.
Ethnic groups: Korean; small Chinese minority.
Religions: Christianity, Buddhism, Shamanism, Confucianism, Chondogyo.
Language: Korean.
Education: *Years compulsory*--9. *Enrollment*--11.5 million. *Attendance*--middle school 99%, high school 95%. *Literacy*--98%.
Health (2000 est.): *Infant mortality rate*--7.85/1,000. *Life expectancy*--men 70.75 yrs.; women 78.5 yrs.
Work force (2004): 22.8 million. *Services*--68%; *mining and manufacturing*--20%; *agriculture*--12%.

Korea's population is one of the most ethnically and linguistically homogenous in the world. Except for a small Chinese community (about 20,000), virtually all Koreans share a common cultural and linguistic heritage. With 48.3 million people, South Korea has one of the world's highest population densities. Major population centers are located in the northwest, southeast, and in the plains south of Seoul-Incheon.

Korea experiences one of the largest rates of emigration with ethnic Koreans residing primarily in China (1.9 million), the United States (1.52 million), Japan (681,000), and the countries of the former Soviet Union (450,000).

The spoken Korean language is very similar to Japanese and differs grammatically from Chinese. Korean also does not use tones. The language is related to Japanese, Mongolian, Hungarian, and other Ural-Altaic languages. Nevertheless, about 90% of all Korean vocabulary has Chinese roots. Chinese ideograms are believed to have been brought into Korea sometime before the second century BC. The learned class spoke Korean, but read and wrote Chinese. The phonetic writing system was invented in the 15th century by Chosun King Sejong to provide a writing system for commoners who could not read classic Chinese. Modern Korean is based almost entirely on the phonetic writing system, although Chinese characters remain in limited use for word clarification. Approximately 1,300 Chinese characters are used in modern Korean. English is taught as a second language in most primary and secondary schools, with unbalanced emphasis on rules of grammar. Chinese and Japanese are widely taught at secondary schools.

Only half of the population actively practices religion. Among this group, Christianity (49%) and Buddhism (47%) comprise Korea's two dominant religions. Though only 3% identified themselves as Confucianists, Korean society remains highly imbued with Confucian values and beliefs. The remaining 1% of the population practice Shamanism (traditional spirit worship) and Chongdogyo ("Heavenly Way"), a traditional religion.

HISTORY

The myth of Korea's foundation by the god-king Tangun in BC 2333 embodies the homogeneity and self-sufficiency valued by the Korean people. Korea experienced many invasions by its larger neighbors in its 2,000 years of recorded history. The country repelled numerous foreign invasions despite domestic strife, in part due to its protected status in the Sino-centric regional political model during Korea's Chosun dynasty (1392-1910). Historical antipathies to foreign influence earned Korea the title of "Hermit Kingdom" in the 19th century.

With declining Chinese power and a weakened domestic posture at the end of the 19th century, Korea was open to Western and Japanese encroachment. In 1910, Japan began a 35-year period of imperial rule over Korea. Memories of Japanese annexation still recall fierce animosity and resentment by older Koreans, as a result of Japan's efforts to supplant the Korean language and culture. Nevertheless, restrictions on Japanese movies, popular music, fashion, etc. have been lifted, and younger Koreans eagerly follow Japanese pop culture.

Japan's surrender to the Allied Powers in 1945, signaling the end of World War II, only further embroiled Korea in foreign rivalries. Division at the 38th Parallel marked the beginning of Soviet and U.S. trusteeship over the North and South, respectively. On August 15, 1948 the Republic of Korea (R.O.K.) was established, with Syngman Rhee as the first President; on September 9, 1948, the Democratic People's Republic of Korea (D.P.R.K.) was established under Kim Il Sung.

On June 25, 1950, North Korean forces invaded South Korea. Led by the U.S., a 16-member coalition undertook the first collective action under United Nations Command (UNC). Shifting battle lines and continuous bombing of the North inflicted a high number of civilian casualties and wrought immense destruction. Following China's entry on behalf of North Korea in 1950, and the stabilization of the front line the following summer, stalemate ensued for the final 2 years of the conflict.

Armistice negotiations, initiated in July 1951, finally concluded on July 27, 1953 at Panmunjom, in the now Demilitarized Zone (DMZ). The resulting Armistice Agreement was signed by the North Korean army, Chinese People's Volunteers, and the U.S.-led and R.O.K.-supported United Nations Command. A peace treaty has never been signed, and the R.O.K. refused to sign the Armistice Agreement.

Domestically, South Korea experienced political turmoil under years of autocratic leadership. Military coups and assassinations characterized the country's first decades. But a vocal civil society emerged that led to strong protests against authoritarian rule. Composed primarily of university students and labor unions, protests reached a climax after Major General Chun Doo Hwan's 1979 military coup and declaration of martial law. A confrontation in Gwangju in 1980 left at least 200 civilians dead but consolidated nationwide support for democracy. In 1987, South Korea was able to hold its first democratic elections in many years.

ECONOMY

Nominal GDP (2002 est.): About $425.0 billion.
GDP growth rate: 2002, 6.0%; 2003, 3.0%.
Per capita GNI (2002 est.): $9,800.
Consumer price index: 2001 avg. increase, 4.1%; 2002, 2.8%.
Natural resources: Limited coal, tungsten, iron ore, limestone, kaolinite, and graphite.
Agriculture, including forestry and fisheries: *Products*--rice, vegetables, fruit. *Arable land*--22% of land area.
Industry: *Types*--Electronics and electrical products, motor vehicles, shipbuilding, mining and manufacturing, petrochemicals, industrial machinery, textiles, footwear.
Trade (2003): *Exports*--$193.8 billion: electronic products (semiconductors, cellular phones, computers), automobiles, machinery and equipment, steel, ships, textiles. *Major markets*--China (including Hong Kong) (20.7%), U.S. (20.2%), European Union (12.8%), Japan (9.3%). *Imports*--$178.8 billion: crude oil, food, machinery and transportation equipment, chemicals and chemical products, base metals and articles. *Major suppliers*--Japan (20.1%), U.S. (13.9%), China (12.3%), European Union (10.6%).

The Republic of Korea's economic growth over the past 30 years has been spectacular. Per capita GNP, only $100 in 1963, exceeded $10,000 in 2003. South Korea is now the United States' seventh-largest trading partner and is the 12th-largest economy in the world.

In the early 1960s, the government of Park Chung Hee instituted sweeping economic policy changes emphasizing exports and labor-intensive light industries, leading to rapid debt-financed industrial expansion. The government carried out a currency reform, strengthened financial institutions, and introduced flexible economic planning. In the 1970s Korea began directing fiscal and financial policies toward promoting heavy and chemical industries, as well as consumer electronics and automobiles. Manufacturing continued to grow rapidly in the 1980s and early 1990s.

In recent years Korea's economy moved away from the centrally planned, government-directed investment model toward a more market-oriented one. Korea bounced back from the 1997-98 crisis with some International Monetary Fund (IMF) assistance, but based largely on extensive financial reforms that restored stability to markets. These economic reforms, pushed by President Kim Dae-jung , helped Korea maintain one of Asia's few expanding economies, with growth rates of 10% in 1999 and 9% in 2000. The slowing global economy and falling exports contributed to slower 3.3% growth in 2001, prompting consumer stimulus measures that led to 6.0% growth in 2002. Consumer over-shopping and rising household debt, along with external factors, slowed growth to below 3% again in 2003. Economic performance in 2004 is expected to improve somewhat, although based largely on vibrant exports.

Economists are concerned that South Korea's economic growth potential has fallen, due to structural problems that are becoming increasingly apparent, along with a rapidly aging population. Foremost among the structural concerns is South Korea's rigid labor market, although the country's underdeveloped financial markets and a general lack of regulatory transparency are also key concerns. Restructuring of Korean conglomerates (chaebols), bank privatization, and creating a more liberalized economy with a mechanism for bankrupt firms to exit the market are also important unfinished reform tasks. Korean industry is increasingly worried about diversion of corporate investment to China.

NORTH-SOUTH TRADE

Two-way trade between the two Koreas has increased from $18.8 million in 1989 to $724 million in 2003. In 2003, South Korea imported $289.2 million worth of goods from North Korea, mostly agro-fisheries and metal products, while shipping $434.9 million worth of goods, mostly humanitarian aid commodities including fertilizer and rice, materials to construct railways and roads, as well as the component parts for processing-on-commission businesses in North Korea. The R.O.K. is North Korea's second-largest trading partner, after China. Numerous ventures by the Hyundai Asan Corporation have contributed to North Korea's economy, including the Mount Keumgang (Diamond Mountain) tourist site. Last year alone, 88,130 visitors traveled by Hyundai-operated passenger ships, and via land routes, as part of this tourism initiative, raising the total number of South Koreans to visit the North to over half a million. Nearly 1,023 North Koreans traveled to South Korea in 2003, mainly for joint sporting events. Hyundai Asan and KOLAND, a Korean Government agency, are co-developing an 800-acre industrial complex in Kaesong, located just north of the DMZ. The year 2003 saw significant progress on reconstructing road and rail links across the DMZ with the Seoul-Kaesong link of the Gyeongui (Western line) scheduled for completion by mid-2004.

U.S.-KOREAN RELATIONS

The United States believes that the question of peace and security on the Korean Peninsula is, first and foremost, a matter for the Korean people to decide.

In the 1954 U.S.-R.O.K. Mutual Defense Treaty, the United States agreed to help the Republic of Korea defend itself against external aggression. In support of this commitment, the United States currently maintains approximately 37,000 service personnel in Korea, including the Army's Second Infantry Division and several Air Force tactical squadrons. To coordinate operations between these units and the 650,000-strong Korean armed forces, a Combined Forces Command (CFC) was established in 1978. The head of the CFC also serves as Commander of the United Nations Command (UNC) and the U.S. Forces in Korea (USFK).

Several aspects of the security relationship are changing as the U.S. moves from a leading to a supporting role. On December 1, 1994, peacetime operational control authority over all South Korean military units still under U.S. operational control was transferred to the South Korean Armed Forces. An agreement has been reached concerning the return of the Yongsan base in the heart of Seoul--as well as a number of other U.S. bases--to the R.O.K. and the relocation of most U.S. forces south of the Han River.

As Korea's economy has developed, trade has become an increasingly important aspect of the U.S.-Korea relationship. The U.S. seeks to improve access to Korea's expanding market and increase investment opportunities for American business. The implementation of structural reforms contained in the IMF's 1998 program for Korea improved access to the Korean market, although a range of serious sectoral and structural barriers still remain. Korean leaders appear determined to successfully manage the complex economic relationship with the United States and to take a more active role in international economic fora as befits Korea's status as a major trading nation.

Statistical Profile: Democratic People's Republic of North Korea

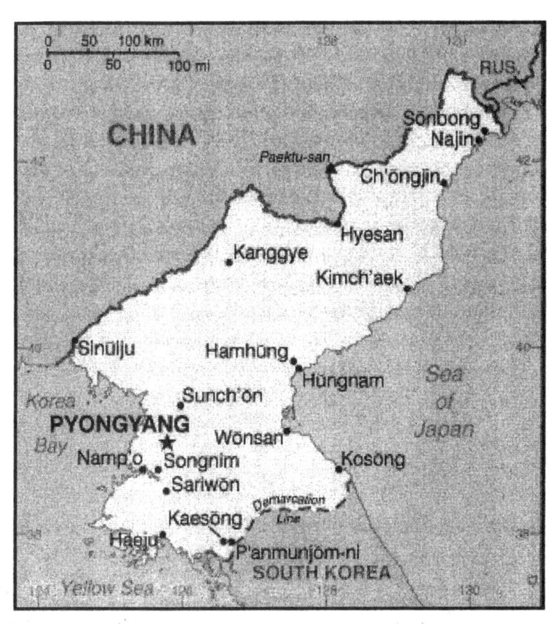

PEOPLE

Nationality: *Noun and adjective*--Korean(s).
Population (2004): 22.7 million.
Annual growth rate: About +0.98%.
Population (2003): 22.5 million.
Annual growth rate: About -0.03%.
Ethnic groups: Korean; small Chinese and Japanese populations.
Religions: Buddhism, Shamanism, Chongdogyo, Christian; religious activities have been virtually nonexistent since 1945.
Language: Korean.
Education: *Years compulsory*--11. *Attendance*--3 million (primary, 1.5 million; secondary, 1.2 million; tertiary, 0.3 million). *Literacy*--99%.
Health (1998): Medical treatment is free; one doctor for every 700 inhabitants; one hospital bed for every 350. *Infant mortality rate*--25/1,000. *Life expectancy*--males 68 yrs., females 74 yrs.

HISTORY

The Korean Peninsula was first populated by peoples of a Tungusic branch of the Ural-Altaic language family, who migrated from the northwestern regions of Asia. Some of these peoples also populated parts of northeast China (Manchuria); Koreans and Manchurians still show physical similarities. Koreans are racially and linguistically homogeneous. Although there are no indigenous minorities in North Korea, there is a small Chinese community (about 50,000) and some 1,800 Japanese wives who accompanied the roughly 93,000 Koreans returning to the North from Japan between 1959 and 1962. Although dialects exist, the Korean spoken throughout the peninsula is mutually comprehensible. In North Korea, the Korean alphabet (hangul) is used exclusively.

Korea's traditional religions are Buddhism and Shamanism. Christian missionaries arrived as early as the 16th century, but it was not until the 19th century that major missionary activity began. Pyongyang was a center of missionary activity, and there was a relatively large Christian population in the north before 1945. Although religious groups exist in North Korea today, the government severely restricts religious activity.

By the first century AD, the Korean Peninsula was divided into the kingdoms of Shilla, Koguryo, and Paekche. In 668 AD, the Shilla kingdom unified the peninsula. The Koryo dynasty--from which Portuguese missionaries in the 16th century derived the Western name "Korea"--succeeded the Shilla kingdom in 935. The Choson dynasty, ruled by members of the Yi clan, supplanted Koryo in 1392 and lasted until Japan annexed Korea in 1910.

Throughout its history, Korea has been invaded, influenced, and fought over by its larger neighbors. Korea was under Mongolian occupation from 1231 until the early 14th century. The unifier of Japan, Hideyoshi, launched major invasions of Korea in 1592 and 1597. When Western powers focused "gunboat" diplomacy on Korea in the mid-19th century, Korea's rulers adopted a closed-door policy, earning Korea the title of "Hermit Kingdom." Though the Choson dynasty recognized China's hegemony in East Asia, Korea was independent until the late 19th century. At that time, China sought to block growing Japanese influence on the Korean Peninsula and Russian pressure for commercial gains there. The competition produced the Sino-Japanese War of 1894-95 and the Russo-Japanese War of 1904-05. Japan emerged victorious from both wars and in 1910 annexed Korea as part of the growing Japanese empire. Japanese colonial administration was characterized by tight control from Tokyo and ruthless efforts to supplant Korean language and culture. Organized Korean resistance during the colonial era was generally unsuccessful, and Japan remained firmly in control of the Peninsula until the end of World War II in 1945. The surrender of Japan in August 1945 led to the immediate division of Korea into two occupation zones, with the U.S. administering the southern half of the peninsula and the U.S.S.R. taking over the area to the north of the 38th parallel. This division was meant to be temporary until the U.S., U.K., Soviet Union, and China could arrange a trusteeship administration.

In December 1945, a conference was convened in Moscow to discuss the future of Korea. A 5-year trusteeship was discussed, and a joint Soviet-American commission was established. The commission met intermittently in Seoul but deadlocked over the issue of establishing a national government. In September 1947, with no solution in sight, the United States submitted the Korean question to the UN

General Assembly. Initial hopes for a unified, independent Korea quickly evaporated as the politics of the Cold War and domestic opposition to the trusteeship plan resulted in the 1948 establishment of two separate nations with diametrically opposed political, economic, and social systems. In 1950, the North launched a massive surprise attack on the South (see, under Foreign Relations, Korean War of 1950-53).

ECONOMY

GDP (2002): $22 billion (purchasing power parity); 30% agriculture, 34% industry, 36% services.
Per capita GDP (2002): $1000 purchasing power parity.
Agriculture: *Products*--rice, corn, potatoes, soybeans, pulses; cattle, pigs, eggs.
Mining and manufacturing: *Types*--military products; machine building, electric power, chemicals; mining (coal, iron ore, etc.), metallurgy; textiles, food processing; tourism.
Trade (2001): *Exports*--$1.044 billion; minerals, metallurgical products, manufactures; textiles, fishery products. *Imports*--$2.042 billion: petroleum, coking coal, machinery and equipment; textiles, grain. *Major partners*--China (39.7%), Thailand (14.6%), Japan (11.2%), Germany (7.6%), R.O.K (6.2%).

North Korea's faltering economy and the breakdown of trade relations with the countries of the former socialist bloc--especially following the fall of communism in eastern Europe and the disintegration of the Soviet Union--have confronted Pyongyang with difficult policy choices. Other centrally planned economies in similar straits have opted for domestic economic reform and liberalization of trade and investment. Despite the introduction of wage and price reforms in 2002, the North Korean leadership seems determined to maintain tight political and ideological control. It has increasingly tolerated markets and a small private sector as the state-run distribution system has deteriorated. Another factor contributing to the economy's poor performance is the disproportionately large percentage of GNP (possibly as much as 25%) that North Korea devotes to the military.

About 80% of North Korea's terrain consists of moderately high mountain ranges and partially forested mountains and hills separated by deep, narrow valleys and small, cultivated plains. The most rugged areas are the north and east coasts. Good harbors are found on the eastern coast. Pyongyang, the capital, near the country's west coast, is located on the Taedong River.

North Korean industry is operating at only a small fraction of capacity due to lack of fuel, spare parts, and other inputs. Agriculture is now 30% of total GNP, even though output has not recovered to early 1990 levels. The infrastructure of the North is generally poor and outdated, and its energy sector has collapsed.

North Korea suffers from chronic food shortages, which were exacerbated by record floods in the summer of 1995 and continued shortages of fertilizer and parts. China and South Korea have responded by making long-term loans on concessional terms to pay for food imports and by direct bilateral food, fertilizer, and energy grants and loans in-kind. International organizations and non-governmental organizations are also providing significant amounts of food. In response to international appeals, the U.S. provided nearly 2 million tons of humanitarian food aid between 1996 and 2003 through the UN World Food Program and through U.S. private voluntary organizations.

TRADE WITH THE U.S.

The United States imposed a total embargo on trade with North Korea in June 1950 when North Korea attacked the South. U.S. law also prohibited financial transactions between the two countries. Since 1989, and most notably in June 2000, the U.S. eased sanctions against North Korea to allow a wide range of exports and imports of U.S. and D.P.R.K. commercial and consumer goods. Imports from North Korea are permitted, subject to an approval process. Direct personal and commercial financial transactions are allowed between U.S. and D.P.R.K. persons. Restrictions on investment also have been eased. Commercial U.S. ships and aircraft carrying U.S. goods are allowed to call at D.P.R.K. ports. The Departments of Commerce and Transportation repealed joint Transportation Order T-2. This order had previously imposed special restrictions on transport to and from North Korea. To date this easing has resulted in little economic activity.

Source: U.S. Department of State (2004)

Chapter References

Fulbright, Leslie. 2004. "East Bay Black Biker Clubs Rev in a New Era. Not All Members Ride Harleys, Some Even Wear Heels." *The Chronicle* (December 19).

Haub, Carl and Machiko Yanagishita. 1994. *1994 World Population Data Sheet.* Washington, DC: Population Reference Bureau.

Hogan, Mary Ann. 1994. "The Joy of Crying." *Los Angeles Times* (February 1):E1+.

Japan Information Center. 1988. *What I Want to Know about Japan.* New York: Consulate General of Japan.

Koehler, Nancy. 1986. "Re-Entry Shock." Pp. 89-94 in *Cross-Cultural Reentry: A Book of Readings* Abilene, TX: Abilene Christian University.

U.S. Department of State. 2004. "Background Note" www.state.gov

Answers

Concept Application

1. Reverse ethnocentrism; re-entry shock — pg. 95; 98
2. Feeling rules — pg. 88
3. Subculture — pg. 100
4. Diffusion — pg. 90
5. Norms — pg. 77

Multiple-Choice

1.	b	pg. 71	11.	c	pg. 84
2.	d	pg. 72	12.	b	pg. 85
3.	d	pg. 74	13.	a	pg. 87
4.	d	pg. 75	14.	c	pg. 88
5.	b	pg. 77	15.	b	pg. 88
6.	b	pg. 77	16.	c	pg. 92
7.	d	pg. 81	17.	b	pg. 92
8.	c	pg. 78	18.	a	pg. 96
9.	c	pg. 81	19.	d	pg. 96
10.	c	pg. 81	20.	c	pg. 100

True/False

1.	F	pg. 85
2.	T	pg. 75
3.	F	pg. 83
4.	F	pg. 76
5.	T	pg. 87
6.	T	pg. 84
7.	F	pg. 90
8.	T	pg. 95
9.	F	pg. 96
10.	T	pg. 92

Chapter 4

Socialization
With Emphasis on Israel, the West Bank, and Gaza

Study Questions

1. Why do Israel, the West Bank, and Gaza receive special attention in a chapter on socialization?

Before 9,11,2001 Americans saw israeli - Palestinian conflicts as Middle East Problem.

2. What is socialization?

A process by which people develop their human capacities and acquire a unique personality and identity and by which culture is passed from generation to generation.

3. What are the basic dynamics underlying the century-long struggle between Palestinian Arabs and Jews? What kinds of issues must be resolved if the peace process is to move forward?

- The land between Jordan River and Mediterranean Sea.
- Both call this land "home"

4. With regard to socialization, what questions would sociologists find interesting to ask about this conflict?

- How do members of a new generation learn about and come to terms with the environment they inherited.
- How is conflict between groups passed down from one generation to another.

5. Distinguish between nature and nurture.

Nature = Human genetic makeup or biological inheritance

Nurture = environment or the interaction experiences that make up every individuals life.

6. How do extreme cases of isolation underscore the importance of socialization? Choose one of the following cases to illustrate: (a) Anna and Isabelle; (b) children orphaned as a result of the Holocaust; (c) Spitz's study of orphanages for children of prison mothers; and (d) the elderly in nursing homes.

–Neglect and lack of socialization can stunt the development of children.

7. On the basis of Anna and Isabelle's case histories, what conclusions did Kingsley Davis reach about the effects of prolonged isolation? What factor did Davis overlook in drawing his conclusions?

- "isolation up to the age of 6, with failure to acquire any form of speech and hence failure to grasp nearly the whole world of cultural meaning."
- Davis overlooked that Isabelle had spent time with her deaf-mute mother and therefore had had some contact.

8. What is the social importance of memory?

- without memory individuals and whole societies would be cut off from the past
  - retains experiences
  - preserves cultural past
  - collective memory (experiences shared by many)

9. What are primary groups? How are they important agents of socialization?

- A social group characterized by face-to-face contact and strong emotional ties among it's members. Ex: Family, teams

10. How does Emile Durkheim define suicide? What are the four types?

Suicide = the severing of relationships
Egoistic – weak ties attaching individual to others in society
Altruistic – person has no life of their own/strive to blend
Anomic – dramatic changes in economic circumstances
Fatalistic – no hope of change

11. What characteristics make a military unit a primary group?

12. What are ingroups and outgroups? What is the sociological significance of ingroups and outgroups?

Ingroup - people identify and feel closely attached, particularly when it is founded on hatred from or opposition towards an outgroup.
outgroup - members of an ingroup feel a sense of separatednedness, opposition or even hatred.

13. How might Durkheim classify the relationships driving Palestinian suicide bombers/martyrs?

Altruistic
- soldiers kill in revenge for their ancestors
- 75% of Palestinians support bombings.

14. Distinguish between the "I" and the "me." How does the "me" develop?

- Me = the social self - part that has learned and internalized societies expectations about appropriate behaviour and appearance
- I = spontaneus, autonomous, creative self, capable of rejecting expectations.

15. How do children come to learn to take the role of others?

1) preparatory
2) play } role taking
3) games

16. According to Charles Horton Cooley's "looking glass-self" theory, how does a sense of self develop?

- product of interaction experiences
- ~~see~~ see themselves reflected in others imagined reactions to their behaviors/appearance.

17. What central concept underlies Piaget's theory of cognitive development? What are the four stages of cognitive development?

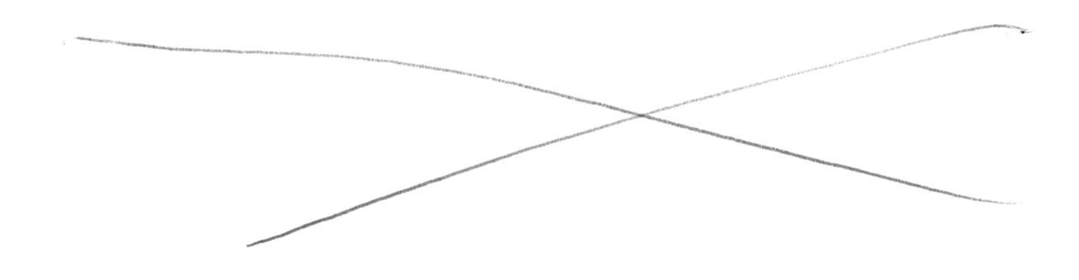

18. What is resocialization? What are the types of resocialization?

The process of discarding values and behaviours unsuited to new circumstances and replacing them with new values and norms...
ex: people marry, change jobs, become parents, change religions
-voluntary or imposed

19. What are total institutions and what mechanisms do total institutions use to resocialize inmates?

-institutions in which people surrender control of their lives (in)voluntarily to a staff and carry out activities with others required
-take away possesions -change appearance -all the same
-limit interaction with outside.

20. Under what conditions are people least likely to resist resocialization?

when they want to be resocialized and when it requires competence not subservience.

21. Expand on the following idea: "Human genetic and social makeup contains considerable potential for change."

Key Concepts

Socialization	pg. 108
Internalization	pg. 108
Nature	pg. 114
Nurture	pg. 114
Engrams	pg. 118
Collective memory	pg. 118
Group	pg. 119
Primary group	pg. 120
Ingroup	pg. 121
Outgroup	pg. 121
Egoistic social relationship	pg. 123

Altruistic social relationship pg. 123
Anomic social relationship pg. 123
Fatalistic social Relationship pg. 123
Reflexive thinking pg. 126
Significant symbol pg. 126
Symbolic gestures pg. 126
Role-taking pg. 127
Play pg. 128
Significant others pg. 128
Games pg. 128
Generalized other pg. 129
Looking-glass self pg. 129

Active adaption pg. 132

Resocialization pg. 134
Total institutions pg. 134

Concept Application

Consider the concepts listed below. Match one or more of the concepts with each scenario. Explain your choices.

- ✓ Collective memory
- ✓ Nature
- ✓ Nurture
- ✓ Resocialization
- ✓ Total institutions
- ✓ Anomic social relationships
- ✓ Altruistic social relationships

Scenario 1 "In 1910, two French surgeons wrote about their successful operation on an 8-year-old boy who had been blind since birth because of cataracts. When the boy's eyes were healed, they removed the bandages, eager to discover how well the child could see. Waving a hand in front of the boy's physically perfect eyes, they asked him what he saw. He replied weakly, 'I don't know.' 'Don't you see it moving? they asked. 'I don't know' was his only reply. The boy's eyes were clearly not following the slowly moving hand. What he saw was only a varying brightness in front of him. He was then allowed to touch the hand. As it began to move; he cried out in a voice of triumph: 'It's moving!' He could feel it move, and even, as he said, 'hear it move,' but he still needed laboriously to learn to *see* it move" (Zajonc 1993:22).

Scenario 2 "Let me say to you, the Palestinians, we are destined to live together on the same soil in the same land. We the soldiers who have returned from battles stained with blood; we who have seen our relatives and friends killed before our eyes; we who have attended their funerals and cannot look into the eyes of their parents; we who have come from a land where parents bury their children; we who have fought against you, the Palestinians" (Rabin 1993:A7).

Scenario 3 "Genetic endowments may set limits for the height or intelligence that individuals can attain but their actual height or intelligence also depends upon how they are raised. The increasing height of the American population over the past several generations reflects the change in nutritional conditions and probably the diminution in childhood illnesses more than a genetic selection" (Lidz 1976:40).

Scenario 4 "Hospitals with hundreds, even thousands of inpatients, maintain schedules aimed at ensuring that every patient receives essential care, and the staff must fit the needs and daily activities of dying patients into the hospital's schedule. They tend to require all patients, whether terminal or not, to give up virtually all personal control over the little things that make up their day-to-day lives. The kinds of personal items that can make a big difference, such as your own pillow from home, are often not allowed. Visits by children may be curtailed, and having a pet stay with a dying person is prohibited. Activities such as walking, eating, bathing, and any physical exercise will proceed according to an established routine" (Anderson 1991:144).

Scenario 5 Some events are experienced by great numbers of people, diverse in interest, age, race, ethnicity, life style and life chances, gender, language, and place, who temporarily become bound together by a historical moment. The January 28, 1986, Space Shuttle *Challenger* disaster was such a moment. Collectively, the country grieved, and not for the first time. Many still vividly remember—and will quickly confess, when the subject comes up—exactly where they were, what they were doing, and how they felt when they heard about the tragedy. The initial shock was perpetuated by the television replays of the *Challenger's* final seconds, the anguished faces of the astronauts' families and other onlookers huddled in disbelief on bleachers at the launch pad, by the news analyses, and then by the official investigation of the Presidential Commission (Vaughn 1996:xi).

Scenario 6 Around 400 volunteers signed up in Tehran to sacrifice their lives in "occupied Islamic countries" on Wednesday night, inspired by a fatwa from a top hardline cleric giving religious backing to suicide missions. Wednesday's registration session was the latest by a group called the Committee for the Commemoration of Martyrs of the Global Islamic Campaign, which says it has enrolled 35,000 volunteers nationwide for possible attacks since last year…"As a Muslim, it is my duty sacrifice my life for oppressed Palestinian children," said Maryam Partovi, 31, a mother of two. A banner hanging over the main entrance quoted Khamenei as saying: "Sacrificing oneself for religion and national interest is the height of honour and bravery."

Applied Research

We learned in Chapter 5 that resocialization is the process of being socialized over again. In particular it is a process of discarding values and behaviors unsuited to new circumstances and replacing them with new, more appropriate values and norms. Identify a resocialization situation—boot camp, first days on a new job, graduation, self-help group, and so on. Use the internet, library research, or personal interviews with someone who has been resocialized to identify critical stages and key events in the resocialization process.

Practice Test: Multiple-Choice Questions

1. The conflict between Israelis and Palestinians has lasted ____________ years.
   a. 100
   b. 50
   c. 25
   d. 5

2. The __________ increased the flow of Jewish refugees to Palestine during World War II.
   a. Six Day War
   b. dropping of the atomic bomb
   c. Nazi Holocaust
   d. *Intifada*

3. In studying the Israeli-Palestinian conflict, sociologists ask all but which one of the following questions?
   a. How do members learn about and come to terms with the environment they have inherited?
   b. How is conflict passed down from one generation to another?
   c. What roles do nature and nurture play in creating a "Palestinian" and an "Israeli" identity?
   d. Why can't we all just get along?

4 Mary was born with blue eyes and blond hair. These characteristics describe
   a. socialization.
   b. internalization.
   c. nature.
   d. nurture.

5. Perhaps the most outstanding feature of the human brain is its flexibility to learn any language. This quality of the brain is an example of
   a. nature.
   b. nurture.
   c. an engram.
   d. a collective memory.

6. The cases of Anna and Isabelle (two cases of extreme isolation) were used to illustrate
   a. the importance of social contact to normal development.
   b. the fact that humans are born with a great learning capacity.
   c. that people are born with preconceived notions about standards of appearance and behavior.
   d. that two-year olds are bothered when rules are violated.

7. __________ is the sociological term for experiences shared and recalled by significant numbers of people.
   a. Group memory
   b. Community memory
   c. Historical memory
   d. Collective memory

8. When author David Grossman asked a group of Palestinian children in a West Bank refugee camp to tell him their birthplace, each replied with the name
   a. Palestine.
   b. Israel.
   c. of a former Arab town.
   d. Jerusalem.

9. Groups that are characterized by face-to-face contact and strong emotional ties among members are
   a. intimate groups.
   b. primary groups.
   c. outgroups.
   d. secondary groups.

10. Sociologists Amith Ben-David and Yoah Lavee's study in which they describe how Israeli family members interacted with one another during the SCUD missile attack, shows that the family
   a. can serve to buffer its members against the effects of negative circumstances, or can exacerbate those effects.
   b. is a supportive and positive influence during a crisis.
   c. increases the stress of a crisis.
   d. becomes divided and tense during a crisis.

11. Which of the following best describes ingroup-outgroup dynamics?
    a. Ingroup members identify with the personal struggles of outgroup members.
    b. The existence of an outgroup can unify attachments to ingroup members.
    c. The presence of an outgroup can unify an ingroup even when the ingroup members are very different from one another.
    d. Because they hate each other, members of an ingroup and outgroup usually find out a lot about one another.

12. The emergence of self depends on our physiological capacity for reflexive thinking which is
    a. the ability to step outside the self and observe and evaluate it from another's viewpoint.
    b. the environment or interaction experiences beginning at birth.
    c. the capacity to learn a spoken language.
    d. the ability to recall memories stored in engrams.

13. The social ties of the chronically ill are characterized as _______________ when family, friends, and other acquaintances avoid interacting with them.
    a. egoistic
    b. altruistic
    c. anomic
    d. fatalistic

14. The ability to step outside the self and observe it from another's viewpoint is
    a. self-awareness.
    b. reflexive thinking.
    c. internalization.
    d. socialization.

15. From the sociological point of view, Mead viewed play and games as important to children's social development because they
    a. give them practice in agility.
    b. increase motor skills.
    c. help develop mental toughness.
    d. allow children to learn and practice taking the role of the other.

16. Palestinian children who pretend to be Israeli soldiers arresting and beating stone throwers are in the _________ stage.
    a. preparatory
    b. play
    c. game
    d. looking-glass self

17. Piaget's ideas about how children develop were inspired by his study of
    a. water snails.
    b. monkeys.
    c. pigeons.
    d. zoo babies.

18. Piaget maintained that learning and reasoning are rooted in
    a. reflexive thinking.
    b. role-playing.
    c. active adaptation.
    d. the looking-glass self.

19. A six-year-old believes that a nail sinks to the bottom of the glass because it is tired. That child is in the __________ stage.
    a. sensorimotor
    b. preoperational
    c. concrete operational
    d. formal operational

20. _________ is the process of discarding values and behaviors unsuited to new circumstances and replacing them with new, more appropriate values and norms.
    a. Reflexive thinking
    b. Role-taking
    c. Ethnocentrism
    d. Resocialization

21. What we know about socialization and resocialization processes suggests that
    a. events unfold in a predictable fashion.
    b. human genetic and social makeup contains considerable potential for change.
    c. no generation can do much to change the problems it inherits from previous generations.
    d. people cannot be resocialized to abandon one way of thinking and behaving for another.

22. ________ wrote *Asylums: Essays on the Social Situation of Mental Patients and Other Inmates*.
    a. Erving Goffman
    b. George Herbert Mead
    c. Charles Horton Cooley
    d. Jean Piaget

23. It is easier to resocialize people if learning a new behavior
    a. is connected to making another party happy.
    b. requires a person to be subservient to another.
    c. leads to a sense of self-worth and competence.
    d. does not take much personal effort.

Practice Test: True/False Questions

T F 1. The West Bank Barrier separates the West Bank from Israel.

T F 2. In the first weeks of life babies are able to babble the sounds needed to speak any language.

T F 3. Memories are stored in engrams (physical traces in the brain) much as films are stored on videocassettes.

T F 4. Military units train their recruits to think of their own personal safety over that of the group.

T F 5. Israeli women are exempt from military service.

T F 6. For suicide bombers/martyrs despair is the factor that best explains his or her willingness to die.

T F 7. Biology is destiny.

T F 8. During the preparatory stage, children practice taking the role of the significant other.

T F 9. Piaget's ideas about how children develop increasingly sophisticated levels of reasoning stems from his study of water snails.

Internet Resources Related to Socialization

- **Human Development**
  http://honolulu.hawaii.edu/intranet/committees/FacDevCom/guidebk/teachtip/teachtip.htm
  As part of its *Faculty Development Teaching Guidebook*, the University of Hawaii includes a section on human development alerting readers to developmental differences across the student population that correspond to chronological age differences. Read "Developmental Tasks of Adolescents, Returning Adults," "Erik H. Erikson's Developmental Stages," and "Malcolm Knowles': Studying the Adult Learner" under the *Human Development* link.

- **George H. Mead and Charles Horton Cooley**
  http://www2.pfeiffer.edu/~lridener/DSS/INDEX.HTML
  A summary of George Herbert Mead's life and ideas is available on the Dead Sociologists Index prepared by Larry R. Ridener. Also check out the life and ideas of Charles Horton Cooley. Mead and Cooley's ideas about the self and its development are critical to the sociological perspective.

Internet Resources Related to Israeli, West Bank, and Gaza

- **News Summaries from Today's Israeli Papers**
  http://www.israelemb.org/useful_links.html
  The Embassy of Israel – Washington, D.C. translates articles from various Israeli dailies including Ma'ariv and HaAretz.

- ***The Jerusalem Report***
  http://www.jrep.com/
  The Jerusalem Report is a bi-weekly publication that "brings the most thorough and in-depth analysis of the controversial issues facing Israel today."

- **Palestinian Central Bureau of Statistics**
  http://www.pcbs.org/
  The Palestinian Central Bureau of Statistics has posted a variety of information on Palestine including population and economic statistics. There is also a list of links to websites such as Palestine Facts, Palestinian Authority and Palestine Colleges and Universities.

Israel: Background Notes

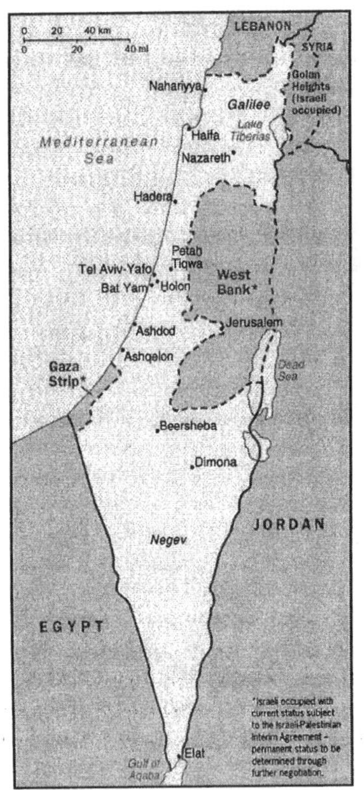

PEOPLE

Population: 6.7 million (November 2003 estimate).
Annual population growth rate: 1.39% (2003 estimate).
Ethnic groups: Jews, 80.1% (slightly less than 5 million), non-Jews (mostly Arab), 19.9% (approximately 1.3 million) (estimates).
Religions: Judaism, Islam, Christianity, Druze.
Languages: Hebrew (official), Arabic (official), English, Russian.
Education: 11 years compulsory. Literacy--95.4% (female 93.6%; male 97.3%).
Health: Infant mortality rate--4.9/1,000, (2002 estimate). Life expectancy at birth--79.02 years; female, 81.19 years, male 76.95 years.
Work force (2.3 million): (1Q 2003) Manufacturing--16.8%; commerce--12.8%; education--2.8%; other business services--12.9%; health and social services--10.2%; community services--4.7%; construction--5.5%; transportation--6.3%; public administration--5.7%; hotels and restaurants--4%; banking and finance--3.4%; agriculture--1.7%; electricity and water-- less than 1%; other--less than 2.2%.

Of the approximately 6.4 million Israelis in 2001, about 5.2 million were counted as Jewish, though some of those are not considered Jewish under Orthodox Jewish law. Since 1989, nearly a million immigrants from the former Soviet Union have arrived in Israel, making this the largest wave of immigration since independence. In addition, almost 50,000 members of the Ethiopian Jewish community have immigrated to Israel, 14,000 of them during the dramatic May 1991 Operation Solomon airlift. Thirty-six percent of Israelis were born outside Israel.

The three broad Jewish groupings are the Ashkenazim, or Jews who trace their ancestry to western, central, and eastern Europe; the Sephardim, who trace their origin to Spain, Portugal, southern Europe, and North Africa; and Eastern or Oriental Jews, who descend from ancient communities in Islamic lands. Of the non-Jewish population, about 73% are Muslims, about 10.5% are Christian, and under 10% are Druze.

Education is compulsory from age 6 to 16 and is free up to age 18. The school system is organized into kindergartens, 6-year primary schools, 3-year junior secondary schools, and 3-year senior secondary schools, after which a comprehensive examination is offered for university admissions. There are seven university-level institutions in Israel, a number of regional colleges, and an Open University program. With a population drawn from more than 100 countries on 5 continents, Israeli society is rich in cultural diversity and artistic creativity. The arts are actively encouraged and supported by the government. The Israeli Philharmonic Orchestra performs throughout the country and frequently tours abroad. The Jerusalem Symphony and the New Israel Opera also tour frequently, as do other musical ensembles.

Almost every municipality has a chamber orchestra or ensemble, many boasting the talents of gifted performers from the countries of the former Soviet Union. Israel has several professional ballet and modern dance companies, and folk dancing, which draws upon the cultural heritage of many immigrant groups, continues to be very popular. There is great public interest in the theater; the repertoire covers the entire range of classical and contemporary drama in translation as well as plays by Israeli authors. Of the three major repertory companies, the most famous, Habimah, was founded in 1917.

Active artist colonies thrive in Safed, Jaffa, and Ein Hod, and Israeli painters and sculptors exhibit works worldwide. Israel boasts more than 120 museums, including the Israel Museum in Jerusalem, which houses the Dead Sea Scrolls along with an extensive collection of regional archaeological artifacts, art, and Jewish religious and folk exhibits. Israelis are avid newspaper readers, with more than 90% of Israeli adults reading a newspaper at least once a week. Major daily papers are in Hebrew; others are in Arabic, English, French, Polish, Yiddish, Russian, Hungarian, and German.

HISTORY

The creation of the State of Israel in 1948 was preceded by more than 50 years of efforts to establish a sovereign nation as a homeland for Jews. These efforts were initiated by Theodore Herzl, founder of the Zionist movement, and were given added impetus by the Balfour Declaration of 1917, which asserted the British Government's support for the creation of a Jewish homeland in Palestine.

In the years following World War I, Palestine became a British Mandate and Jewish immigration steadily increased, as did violence between Palestine's Jewish and Arab communities. Mounting British efforts to restrict this immigration were countered by international support for Jewish national aspirations following the near-extermination of European Jewry by the Nazis during World War II. This support led to the 1947 UN partition plan, which would have divided Palestine into separate Jewish and Arab states, with Jerusalem under UN administration.

On May 14, 1948, soon after the British quit Palestine, the State of Israel was proclaimed and was immediately invaded by armies from neighboring Arab states, which rejected the UN partition plan. This conflict, Israel's War of Independence, was concluded by armistice agreements between Israel, Egypt, Jordan, Lebanon, and Syria in 1949 and resulted in a 50% increase in Israeli territory.

In 1956, French, British, and Israeli forces engaged Egypt in response to its nationalization of the Suez Canal and blockade of the Straits of Tiran. Israeli forces withdrew in March 1957, after the United Nations established the UN Emergency Force (UNEF) in the Gaza Strip and Sinai. This war resulted in no territorial shifts and was followed by several years of terrorist incidents and retaliatory acts across Israel's borders.

In June 1967, Israeli forces struck targets in Egypt, Jordan, and Syria in response to Egyptian President Nasser's ordered withdrawal of UN peacekeepers from the Sinai Peninsula and the buildup of Arab armies along Israel's borders. After 6 days, all parties agreed to a cease-fire, under which Israel retained control of the Sinai Peninsula, the Golan Heights, the Gaza Strip, the formerly Jordanian-controlled West Bank of the Jordan River, and East Jerusalem. On November 22, 1967, the Security Council adopted Resolution 242, the "land for peace" formula, which called for the establishment of a just and lasting peace based on Israeli withdrawal from territories occupied in 1967 in return for the end of all states of belligerency, respect for the sovereignty of all states in the area, and the right to live in peace within secure, recognized boundaries.

The following years were marked by continuing violence across the Suez Canal, punctuated by the 1969-70 war of attrition. On October 6, 1973--Yom Kippur (the Jewish Day of Atonement), the armies of Syria and Egypt launched an attack against Israel. Although the Egyptians and Syrians initially made significant advances, Israel was able to push the invading armies back beyond the 1967 cease-fire lines by the time the United States and the Soviet Union helped bring an end to the fighting. In the UN Security Council, the United States supported Resolution 338, which reaffirmed Resolution 242 as the framework for peace and called for peace negotiations between the parties.

In the years that followed, sporadic clashes continued along the cease-fire lines but guided by the U.S., Egypt and Israel continued negotiations. In November 1977, Egyptian President Anwar Sadat made a historic visit to Jerusalem, which opened the door for the 1978 Israeli-Egyptian peace summit convened at Camp David by President Carter. These negotiations led to a 1979 peace treaty between Israel and Egypt, pursuant to which Israel withdrew from the Sinai in 1982, signed by President Sadat of Egypt and Prime Minister Menahem Begin of Israel.

In the years following the 1948 war, Israel's border with Lebanon was quiet relative to its borders with other neighbors. After the expulsion of Palestinian fighters from Jordan in 1970 and their influx into southern Lebanon, however, hostilities along Israel's northern border increased and Israeli forces crossed into Lebanon. After passage of Security Council Resolution 425, calling for Israeli withdrawal and the creation of the UN Interim Force in Lebanon peacekeeping force (UNIFIL), Israel withdrew its troops. In June 1982, following a series of cross-border terrorist attacks and the attempted assassination of the Israeli Ambassador to the U.K., Israel invaded Lebanon to fight the forces of Yasser Arafat's Palestine Liberation Organization (PLO). The PLO withdrew its forces from Lebanon in August 1982. Israel, having failed to finalize an agreement with Lebanon, withdrew most of its troops in June 1985 save for a residual force which remained in southern Lebanon to act as a buffer against attacks on northern Israel. These remaining forces were completely withdrawn in May 2000 behind a UN-brokered delineation of the Israel-Lebanon border (the Blue Line). Hizballah forces in Southern Lebanon continued to attack Israeli positions south of the Blue Line in the Sheba Farms/Har Dov area of the Golan Heights.

The victory of the U.S.-led coalition in the Persian Gulf War of 1991 opened new possibilities for regional peace. In October 1991, the United States and the Soviet Union convened the Madrid Conference, in which Israeli, Lebanese, Jordanian, Syrian, and Palestinian leaders laid the foundations for ongoing negotiations designed to bring peace and economic development to the region. Within this framework, Israel and the PLO signed a Declaration of Principles on September 13, 1993, which established an ambitious set of objectives relating to a transfer of authority from Israel to an interim Palestinian authority. Israel and the PLO subsequently signed the Gaza-Jericho Agreement on May 4, 1994, and the Agreement on Preparatory Transfer of Powers and Responsibilities on August 29, 1994, which began the process of transferring authority from Israel to the Palestinians.

On October 26, 1994, Israel and Jordan signed a historic peace treaty, witnessed by President Clinton. This was followed by Israeli Prime Minister Rabin and PLO Chairman Arafat's signing of the historic Israeli-Palestinian Interim Agreement on September 28, 1995. This accord, which incorporated and superseded previous agreements, broadened Palestinian self-government and provided for cooperation between Israel and the Palestinians in several areas.

Israeli Prime Minister Yitzhak Rabin was assassinated on November 4, 1995, by a right-wing Jewish radical, bringing the increasingly bitter national debate over the peace process to a climax. Subsequent Israeli governments continued to negotiate with the PLO resulting in additional agreements, including the Wye River and the Sharm el-Sheikh memoranda.

A summit hosted by President Clinton at Camp David in July 2000 to address permanent status issues--including the status of Jerusalem, Palestinian refugees, Israeli settlements in the West Bank and Gaza, final security arrangements, borders, and relations and cooperation with neighboring states--failed to produce an agreement.

Following the failed talks, widespread violence broke out in Israel, the West Bank, and Gaza in September 2000. In April 2001 the Sharm el-Sheikh Fact Finding Committee, commissioned by the October 2000 Middle East Peace Summit and chaired by former U.S. Senator George Mitchell, submitted its report, which recommended an immediate end to the violence followed by confidence-building measures and a resumption of security cooperation and peace negotiations. The United States has worked intensively to help bring an end to the violence between Israelis and Palestinians and bring about the implementation of the recommendations of the Mitchell Committee as a bridge back to political negotiations. In April 2003, the Quartet (the U.S., U.N., E.U., and the Russian Federation) announced the "roadmap," a performance-based plan to bring about two states, Israel and a democratic, viable Palestine, living side by side in peace and security. Both the Israelis and Palestinians have affirmed their commitment to the roadmap, but continuing Israeli-Palestinian violence has led to a continuing crisis of confidence between the two sides.

Despite the promising developments of spring 2003, violence continued and in September 2003 the first Palestinian Prime Minister, Mahmoud Abbas (Abu Mazin), resigned after failing to win true authority to restore law and order, fight terror, and reform Palestinian institutions. In response to the deadlock, in the winter of 2003-2004 Prime Minister Sharon put forward his Gaza disengagement plan, proposing the withdrawal of Israeli settlements from Gaza as well as parts of the northern West Bank. President Bush endorsed this initiative in an exchange of letters with Prime Minister Sharon on April 14, 2004, viewing the Gaza disengagement initiative as an opportunity to move towards implementation of the two-state vision and begin the development of Palestinian institutions. The Quartet endorsed the initiative in a meeting in May 2004 and since then the United States has been working intensively with the parties to the conflict, regional partners, and the broad international community to make Gaza disengagement a reality.

ECONOMY

Israel has a diversified, technologically advanced economy with substantial but decreasing government ownership and a strong high-tech sector. The major industrial sectors include high-technology electronic and biomedical equipment, metal products, processed foods, chemicals, and transport equipment. Israel possesses a substantial service sector and is one of the world's centers for diamond cutting and polishing. It also is a world leader in software development and, prior to the violence that began in September 2000, was a major tourist destination.

Israel's strong commitment to economic development and its talented work force led to economic growth rates during the nation's first two decades that frequently exceeded 10% annually. The years after the 1973 Yom Kippur War were a lost decade economically, as growth stalled and inflation reached triple-digit levels. The successful economic stabilization plan implemented in 1985 and the subsequent introduction of market-oriented structural reforms reinvigorated the economy and paved the way for rapid growth in the 1990s.

A wave of Jewish immigration beginning in 1989, predominantly from the countries of the former U.S.S.R., brought nearly a million new citizens to Israel. These new immigrants, many of them highly educated, now constitute some 13% of Israel's 6.7 million inhabitants. Their successful absorption into Israeli society and its labor force forms a remarkable chapter in Israeli history. The skills brought by the new immigrants and their added demand as consumers gave the Israeli economy a strong upward push and in the 1990s they played a key role in the ongoing development of Israel's high-tech sector. During the 1990s, progress in the Middle East peace process, beginning with the Madrid Conference of 1991, helped to reduce Israel's economic isolation from its neighbors and opened up new markets to Israeli exporters farther afield. The peace process stimulated an unprecedented inflow of foreign investment in Israel, and provided a substantial boost to economic growth in the region over the last decade. The onset of the intifada beginning at the end of September of 2000, the downturn in the high-tech sector and Nasdaq crisis, and the slowdown of the global economy--particularly the U.S. economy--have all significantly affected the Israeli economy during the past three years.

Israeli companies, particularly in the high-tech area, have in the past enjoyed considerable success raising money on Wall Street and other world financial markets; Israel ranks second to Canada among foreign countries in the number of its companies listed on U.S. stock exchanges. Israel's tech market is very developed, and in spite of the pause in the industry's growth, the high-tech sector is likely to be the major driver of the Israeli economy. Almost half of Israel's exports are high tech. Most leading players, including Intel, IBM, and Cisco have a presence in Israel, and it is worth noting that even during the downturn in the macroeconomic situation in Israel these large players as well as others did not withdraw from the Israeli market.

Growth was an exceptional 6.2% in 2000, due in part to a number of one-time high tech acquisitions and investments. This exceptional year was followed by two years of negative growth of –0.9% and –1%, respectively, in 2001 and 2002. As a result of the security situation, and associated downturn in the economy, there has been a significant rise in unemployment and wage erosion. This led to a decline in private consumption in 2002, the first time that there had been negative private consumption since the early 1980's. The economy grew marginally in 2003 at a rate of 1.2%. The change in the geopolitical situation as a result of the successful completion of the War in Iraq, combined with the potential for some progress in the political situation, as well as the approval of a GOI economic recovery plan, and approval of U.S. loan guarantees, are likely to have positive effects on the economy.

The United States is Israel's largest trading partner. In 2002, two-way trade totaled some $19.66 billion, and Israel had approximately a $5.88 billion trade surplus with the U.S. The principal U.S. exports to Israel include civilian aircraft parts, telecommunications equipment, semiconductors, civilian aircraft, electrical apparatus, and computer accessories. Israel's chief exports to the U.S. include diamonds, pharmaceutical preparations, telecommunications equipment, medicinal equipment, electrical apparatus, and cotton apparel. The two countries signed a free trade agreement (FTA) in 1985 that progressively eliminated tariffs on most goods traded between the two countries over the following 10 years. An agricultural trade accord signed in November 1996 addressed the remaining goods not covered in the FTA but has not entirely erased barriers to trade in the agricultural sector. Israel also has trade and cooperation agreements in place with the European Union, Canada, Mexico, and other countries. Best prospect industry sectors in Israel for U.S. exporters are electricity and gas equipment, defense equipment, medical instruments and disposable products, industrial chemicals, telecommunication equipment, electronic components, building materials/construction industries (DIY and infrastructure), safety and security equipment and services, non-prescription drugs, travel and tourism services, and computer software.

FOREIGN RELATIONS

In addition to seeking an end to hostilities with Arab forces, against which it has fought five wars since 1948, Israel has given high priority to gaining wide acceptance as a sovereign state with an important international role. Before 1967, it had established diplomatic relations with a majority of the world's

nations, except for the Arab states and most other Muslim countries. The Soviet Union and the communist states of eastern Europe (except Romania) broke diplomatic relations with Israel during the 1967 war, but those relations were restored by 1991.
Today, Israel has diplomatic relations with 161 states. Following the signing of the Israel-PLO Declaration of Principles on September 13, 1993, Israel established or renewed diplomatic relations with 35 countries. Most important are its ties with Arab states. Israel has full diplomatic relations with Egypt and Jordan.

On October 1, 1994, the Gulf States publicly announced their support for a review of the Arab boycott, in effect abolishing the secondary and tertiary boycotts against Israel. Israel has diplomatic relations with nine non-Arab Muslim states and with 32 of the 43 Sub-Saharan states that are not members of the Arab League. Israel established relations with China and India in 1992 and with the Holy See in 1993.

U.S.-ISRAELI RELATIONS

Commitment to Israel's security and well being has been a cornerstone of U.S. policy in the Middle East since Israel's creation in 1948, in which the United States played a key supporting role. Israel and the United States are bound closely by historic and cultural ties as well as by mutual interests. Continuing U.S. economic and security assistance to Israel acknowledges these ties and signals U.S. commitment. The broad issues of Arab-Israeli peace have been a major focus in the U.S.-Israeli relationship. U.S. efforts to reach a Middle East peace settlement are based on UN Security Council Resolutions 242 and 338 and have been based on the premise that as Israel takes calculated risks for peace, the United States will help minimize those risks.

UNSC resolutions provided the basis for cease-fire and disengagement agreements concerning the Sinai and the Golan Heights between Israel, Egypt, and Syria and for promoting the Camp David accords and the Egyptian-Israeli Peace Treaty.

The landmark October 1991 Madrid conference also recognized the importance of Security Council Resolutions 242 and 338 in resolving regional disputes, and brought together for the first time Israel, the Palestinians, and the neighboring Arab countries, launching a series of direct bilateral and multilateral negotiations. These talks were designed to finally resolve outstanding security, border, and other issues between the parties while providing a basis for mutual cooperation on issues of general concern, including the status of refugees, arms control and regional security, water and environmental concerns, and economic development.

On a bilateral level, relations between the United States and Israel have been strengthened in recent years by the establishment of cooperative institutions in many fields. Bilateral foundations in the fields of science and technology include the Binational Science Foundation and the Binational Agricultural Research and Development Foundation. The U.S.-Israeli Education Foundation sponsors educational and cultural programs.

In addition, the Joint Economic Development Group maintains a high-level dialogue on economic issues. In early 1993, the United States and Israel agreed to establish a Joint Science and Technology Commission. In 1996, reflecting heightened concern about terrorism, the United States and Israel established a Joint Counterterrorism Group designed to enhance cooperation in fighting terrorism.

Source: U.S. Department of State (2004)

Gaza Strip: Background Notes

BACKGROUND

The Israel-PLO Declaration of Principles on Interim Self-Government Arrangements (the DOP), signed in Washington on 13 September 1993, provided for a transitional period not exceeding five years of Palestinian interim self-government in the Gaza Strip and the West Bank. Under the DOP, Israel agreed to transfer certain powers and responsibilities to the Palestinian Authority, which includes the Palestinian Legislative Council elected in January 1996, as part of the interim self-governing arrangements in the West Bank and Gaza Strip. A transfer of powers and responsibilities for the Gaza Strip and Jericho took place pursuant to the Israel-PLO 4 May 1994 Cairo Agreement on the Gaza Strip and the Jericho Area and in additional areas of the West Bank pursuant to the Israel-PLO 28 September 1995 Interim Agreement, the Israel-PLO 15 January 1997 Protocol Concerning Redeployment in Hebron, the Israel-PLO 23 October 1998 Wye River Memorandum, and the 4 September 1999 Sharm el-Sheikh Agreement. The DOP provides that Israel will retain responsibility during the transitional period for external security and for internal security and public order of settlements and Israeli citizens. Permanent status is to be determined through direct negotiations, which resumed in September 1999 after a three-year hiatus. An intifada broke out in September 2000; the resulting widespread violence in the West Bank and Gaza Strip, Israel's military response, and instability in the Palestinian Authority are undermining progress toward a permanent settlement.

AREA

Total area: 360 sq km
Land area: 360 sq km
Water area: 0 sq km

Comparative area: Slightly more than twice the size of Washington, DC
Natural resources: Arable land, natural gas

ENVIRONMENT

Current issues: Desertification; salination of fresh water; sewage treatment; water-borne disease; soil degradation

PEOPLE

Population: 1,324,991
note: in addition, there are more than 5,000 Israeli settlers in the Gaza Strip (July 2004 est.)

Age structure:
0-14 years: 49% (male 332,582; female 316,606)
15-64 years: 48.3% (male 326,450; female 314,098)
65 years and over: 2.7% (male 14,847; female 20,408) (2004 est.)

Population growth rate: 3.83% (2004 est.)

Birth rate: 40.62 births/1,000 population (2004 est.)

Death rate: 3.95 deaths/1,000 population (2004 est.)

Net migration rate: 1.6 migrant(s)/1,000 population (2004 est.)

Sex ratio:
at birth: 1.05 male(s)/female
under 15 years: 1.05 male(s)/female
15-64 years: 1.04 male(s)/female
65 years and over: 0.73 male(s)/female
total population: 1.04 male(s)/female (2004 est.)

Infant mortality rate: 23.54 deaths/1,000 live births (2004 est.)

Life expectancy at birth:
total population: 71.59 years
male: 70.31 years
female: 72.94 years (2004 est.)

Total fertility rate: 6.04 children born/woman (2004 est.)

ECONOMY

Overview: Economic output in the Gaza Strip - under the responsibility of the Palestinian Authority since the Cairo Agreement of May 1994 - declined by about one-third between 1992 and 1996. The downturn was largely the result of Israeli closure policies - the imposition of generalized border closures in response to security incidents in Israel - which disrupted previously established labor and commodity market relationships between Israel and the WBGS (West Bank and Gaza Strip). The most serious negative social effect of this downturn was the emergence of high unemployment; unemployment in the WBGS during the 1980s was generally under 5%; by 1995 it had risen to over 20%. Israel's use of comprehensive closures decreased during the next few years and, in 1998, Israel implemented new policies to reduce the impact of closures and other security procedures on the movement of Palestinian goods and labor. These changes fueled an almost three-year-long economic recovery in the West Bank and Gaza Strip; real GDP grew by 5% in 1998 and 6% in 1999. Recovery was upended in the last quarter of 2000 with the outbreak of violence, triggering tight Israeli closures of Palestinian self-rule areas and a

severe disruption of trade and labor movements. In 2001, and even more severely in 2002, Israeli military measures in Palestinian Authority areas resulted in the destruction of capital plant and administrative structure, widespread business closures, and a sharp drop in GDP. Including the West Bank, the UN estimates that more than 100,000 Palestinians out of the 125,000 who used to work in Israel, in Israeli settlements, or in joint industrial zones have lost their jobs. In addition, about 80,000 Palestinian workers inside the Territories are losing their jobs. International aid of $2 billion in 2001-02 to the West Bank and Gaza Strip prevented the complete collapse of the economy and allowed Finance Minister Salam Fayyad to implement several financial and economic reforms. Budgetary support, however, was not as forthcoming in 2003.
GDP - per capita: purchasing power parity - $600 (2003 est.)

GDP - composition by sector: *agriculture:* 9%
industry: 28%
services: 63% (includes West Bank) (2002 est.)

Source: The World Factbook, Gaza Strip (2004), http://www.odci.gov/cia/publications/factbook/

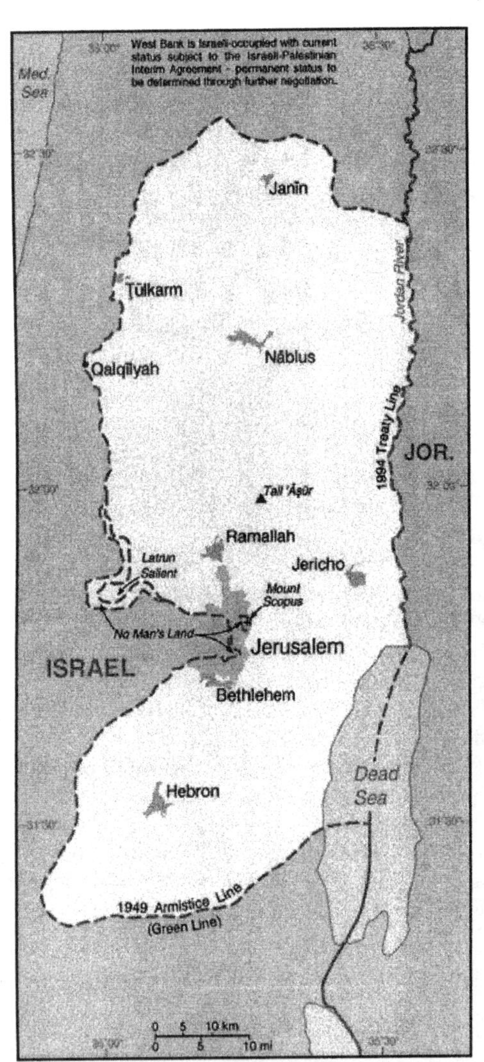

West Bank: Background Notes

BACKGROUND

The Israel-PLO Declaration of Principles on Interim Self-Government Arrangements (the DOP), signed in Washington on 13 September 1993, provided for a transitional period not exceeding five years of Palestinian interim self-government in the Gaza Strip and the West Bank. Under the DOP, Israel agreed to transfer certain powers and responsibilities to the Palestinian Authority, which includes the Palestinian Legislative Council elected in January 1996, as part of interim self-governing arrangements in the West Bank and Gaza Strip. A transfer of powers and responsibilities for the Gaza Strip and Jericho took place pursuant to the Israel-PLO 4 May 1994 Cairo Agreement on the Gaza Strip and the Jericho Area and in additional areas of the West Bank pursuant to the Israel-PLO 28 September 1995 Interim Agreement, the Israel-PLO 15 January 1997 Protocol Concerning Redeployment in Hebron, the Israel-PLO 23 October 1998 Wye River Memorandum, and the 4 September 1999 Sharm el-Sheikh Agreement. The DOP provides that Israel will retain responsibility during the transitional period for external security and for internal security and public order of settlements and Israeli citizens. Permanent status is to be determined through direct negotiations, which resumed in September 1999 after a three-year hiatus. An intifada broke out in September 2000; the resulting widespread violence in the West Bank and

Gaza Strip, Israel's military response, and instability in the Palestinian Authority are undermining progress toward a permanent settlement.

AREA

Total area: 5,860 sq km
Land area: 5,640 sq km
Water area: 220 sq km
n*ote:* includes West Bank, Latrun Salient, and the northwest quarter of the Dead Sea, but excludes Mt. Scopus; East Jerusalem and Jerusalem No Man's Land are also included only as a means of depicting the entire area occupied by Israel in 1967
Comparative area: Slightly smaller than Delaware
Natural resources: Arable land

ENVIRONMENT

Current issues: Adequacy of fresh water supply; sewage treatment

PEOPLE

Population: 2,311,204
note: in addition, there are about 187,000 Israeli settlers in the West Bank and fewer than 177,000 in East Jerusalem (July 2004 est.)

Age structure:
0-14 years: 43.8% (male 518,470; female 493,531)
15-64 years: 52.8% (male 623,785; female 595,376)
65 years and over: 3.5% (male 34,226; female 45,816) (2004 est.)

Population growth rate: 3.21% (2004 est.)

Birth rate: 33.21 births/1,000 population (2004 est.)

Death rate: 4.07 deaths/1,000 population (2004 est.)

Net migration rate: 2.98 migrant(s)/1,000 population (2004 est.)

Sex ratio:
at birth: 1.06 male(s)/female
under 15 years: 1.05 male(s)/female
15-64 years: 1.05 male(s)/female
65 years and over: 0.75 male(s)/female
total population: 1.04 male(s)/female (2004 est.)

Infant mortality rate: 20.16 deaths/1,000 live births (2004 est.)

Life expectancy at birth:
total population: 72.88 years
male: 71.14 years
female: 74.72 years (2004 est.)

Total fertility rate: 4.52 children born/woman (2004 est.)

ECONOMY

Overview: Real per capita GDP for the West Bank and Gaza Strip (WBGS) declined by about one-third between 1992 and 1996 due to the combined effect of falling aggregate incomes and rapid population growth. The downturn in economic activity was largely the result of Israeli closure policies - the imposition of border closures in response to security incidents in Israel - which disrupted labor and commodity market relationships between Israel and the WBGS. The most serious social effect of this downturn was rising unemployment, which in the WBGS during the 1980s was generally under 5%; by 1995 it had risen to over 20%. Israel's use of comprehensive closures during the next three years decreased and, in 1998, Israel implemented new policies to reduce the impact of closures and other security procedures on the movement of Palestinian goods and labor. These changes fueled an almost three-year-long economic recovery in the West Bank and Gaza Strip; real GDP grew by 5% in 1998 and 6% in 1999. Recovery was upended in the last quarter of 2000 with the outbreak of violence, which triggered tight Israeli closures of Palestinian self-rule areas and severely disrupted trade and labor movements. In 2001, and even more severely in 2002, Israeli military measures in Palestinian Authority areas resulted in the destruction of much capital plant and administrative structure, widespread business closures, and a sharp drop in GDP. Including Gaza Strip, the UN estimates that more than 100,000 Palestinians out of the 125,000 who used to work in Israel, in Israeli settlements, or in joint industrial zones have lost their jobs. In addition, about 80,000 Palestinian workers inside the Territories are losing their jobs. International aid of $2 billion in 2001-02 to the West Bank and Gaza Strip prevented the complete collapse of the economy. In 2004, on-going border issues and the death of Yasser Arafat continued to complicate the economic situation.
GDP - per capita: purchasing power parity - $800 (2002 est.)

GDP - composition by sector: *agriculture:* 9%
industry: 28%
services: 63%
note: includes Gaza Strip (1999 est.)

Sources: The World Factbook, the West Bank (2004)
http://www.odci.gov/cia/publications/factbook/

Chapter References

Hafezi, Parisa. 2005. "Iranian Hardliners Register as Suicide Bombers." Reuters Foundation. http://www.alertnet.org/thenews.

Haub, Carl and Machiko Yanagishita. 1994. *1994 World Population Data Sheet.* Washington, DC: Population Reference Bureau.

Lidz, Theodore. 1976. *The Person: His and Her Development Throughout the Life Cycle.* New York: Basic Books.

Mura, David. 1996. *Where Body Meets Memory.* New York: Anchor.

Rabin, Yitzhak. 1993. "Making a New Middle East. 'Shalom, Salaam, Peace.' Views of Three Leaders." *Los Angeles Times* (September 14):A7.

Vaughn, Diane. 1996. *The Challenger Launch*. Chicago: University of Chicago.

Zajonc, Arthur. 1993. "Seeing the Light." *Los Angeles Times Magazine* (July 25):22-25.

Answers

Concept Application

1. Nature; Nurture; Resocialization pg. 114; 134
2. Collective memory pg. 118
3. Nature and Nurture pg. 114
4. Resocialization, total institutions pg. 134
5. Collective memory pg. 118
6. Altruistic social relationship pg. 123

Multiple-Choice

1.	a	pg. 106	13.	a	pg. 123
2.	c	pg. 109	14.	b	pg. 126
3.	d	pg. 114	15.	d	pg. 127
4.	c	pg. 114	16.	b	pg. 128
5.	a	pg. 115	17.	a	pg. 132
6.	a	pg. 115	18.	c	pg. 132
7.	d	pg. 118	19.	b	pg. 133
8.	c	pg. 118	20.	d	pg. 134
9.	b	pg. 120	21.	b	pg. 135
10.	a	pg. 120	22.	a	pg. 134
11.	c	pg. 121	23.	c	pg. 135
12.	a	pg. 126			

True/False

1.	T	pg. 113
2.	T	pg. 115
3.	F	pg. 118
4.	F	pg. 120
5.	F	pg. 120
6.	F	pg. 124
7.	F	pg. 136
8.	F	pg. 127
9.	T	pg. 132

Chapter 5

Social Interaction and the Social Construction of Reality
With Emphasis on the Democratic Republic of the Congo

Study Questions

1. How is the Democratic Republic of the Congo related to issues of social interaction and the social construction of reality?

2. What is social interaction? How do sociologists approach social interaction?

Everyday events in which 2+ people communicate and respond through language/symbolic gestures to affect one another's behaviour and thinking.
- seek to understand/explain forces of context/content.

3. How did Durkheim define the division of labor? How is the division of labor related to global interdependence?

increase in pop size/density increases resource demand which stimulates people to find more efficient methods, advancing society.
- work that is broken down into specialized tasks.

4. Distinguish between organic and mechanical solidarity.

organic - social order based on interdependence/cooperation among people performing a wide range of specialized tasks
mechanical - social order and cohesion based on a common conscience or uniform thinking/behaviour.

5. What kinds of disruptive events break down the abilities of individuals to connect with one another in meaningful ways through their labor?

6. Use the case of the Congo to give an example of each of the five disruptive events to the division of labor. (An example may be used more than once.)

7. How is the activation and spread of HIV connected to the unprecedented mixing of people from all over the world?

8. Who is responsible for triggering and transmitting HIV? Explain your answer.

9. How can people interact smoothly with people they know nothing about? Explain.

10. What is a social status? How is it related to social structure?

- A position in a social structure.
- social structure - 2+ people occupying social statuses and enacting roles.

11. How are the concepts of status, role set, rights, and obligations related?

Role set - an array of roles
rights - behaviours that a person assuming a role can demand / expect from others.
obligation - relationship / behaviour that the person enacting a role must assume towards others in a particular status.

12. Does the idea of role imply totally predictable behavior? Explain.

13. What are the broad differences between practitioner-patient roles in Africa and in the United States?

14. What is impression management? What interaction dilemmas are associated with impression management?

15. How does the concept of impression management help us to understand the dilemmas health leaders face in the United States and the Congo in convincing people to use condoms?

16. What is the difference between back stage and front stage? Use these concepts to analyze blood banks and their handling of HIV.

17. What sociological concepts would you draw upon to analyze the content of interaction?

18. People usually attribute cause to either dispositional traits or situational factors. What is the difference between dispositional traits and structural factors? Give an example of each.

19. What problems are associated with using dispositional traits to explain the cause of AIDS and to diagnose AIDS cases?

20. What must take place before we can truly understand the cause of HIV and AIDS?

21. Give an example of how research decisions about labeling and categorizing HIV-risk shapes our understanding of this condition.

22. Explain the following statement: "Television uses an image-oriented format." What are the problems associated with such a format?

Key Concepts

Social interaction	pg. 142
Context of interaction	pg. 142
Division of labor	pg. 142
Solidarity	pg. 144
Mechanical	pg. 144
Organic	pg. 146
Content of interaction	pg. 142
Social structure	pg. 153
Social Status	pg. 153
Role	pg. 153
Role set	pg. 154
Rights	pg. 154
Obligation	pg. 154
Role strain	pg. 154
Role conflict	pg. 154
Dramaturgical model	pg. 156
Impression management	pg. 157
Front stage	pg. 159
Back stage	pg. 159
Attribution	pg. 163
Thomas theorem	pg. 163
Dispositional traits	pg. 163
Situational traits	pg. 163
Scapegoat	pg. 165

Concept Application

Consider the concepts listed below. Match one or more of the concepts with each scenario. Explain your choices.

- ✓ Dispositional traits
- ✓ Front stage
- ✓ Impression management
- ✓ Role Strain
- ✓ Situational factors
- ✓ Social interactions

Scenario 1 "Ten minutes after William Andrews succumbed to the poisonous concoction injected into his arm, Dr. Robert Jones performed a task from which, he said, he would never quite recover. He entered the chamber of death, checked the condemned man's vital signs and confirmed that he was, in fact, dead.

The medical director for the Utah State Prison system did not witness the July, 1992, execution. But his limited role so troubled him that he decided never again to have anything to do with a state-ordered killing.

'It was much more stressful, much more disconcerting than I though it would be,' Jones says. 'I literally slept for a whole day afterward, and I thought, 'That's an experience in life that you don't want to have to go through again.... Physicians usually try to preserve life, not end it.'

As a prison doctor, Jones sits at the uncomfortable intersection of medicine and criminal justice. His dilemma highlights an ethical debate that is raging in the medical community: Should doctors, who take the Hippocratic oath not to harm their patients, take part in carrying out the death penalty? When state laws and regulations require physicians to be present at executions—as in California, where doctors watch the heart monitor that charts the prisoner's final moments in the gas chamber—should the physician comply?" (Stolberg 1994:E1)

Scenario 2 "Janet's sister, Pam, and brother, Nicholas, along with their own spouses and children, had a hard time understanding what was happening to their mother. It took them longer than it took Janet to catch on, because their mother managed to do a superb job of keeping up appearances during the quarter of an hour or so each week when they spoke with her on the phone. And because they didn't want anything major to be the matter either, they weren't able to take Janet's worrying seriously for quite a long time" (Nelson and Nelson 1996:44).

Scenario 3 "Almost everybody, at some point in life, will avoid uncomfortable truths, 'edit' their own memories, mislead others, and even sometimes tell out-and-out falsehoods. And almost everybody feels uncomfortable about lying repeatedly. As Barbara wrestles with this problem, she has put herself in her dad's shoes and acknowledged that she would feel very uncomfortable if it turned out that someone was lying to her. Even so, she also knows from experience that the price of avoiding a lie can sometimes be just as high as the price of telling one.

While people will agree that one ought to tell the truth whenever possible, it's not so easy to say precisely why that's so. To understand better whether and when it's morally okay to break the rule against lying, it's necessary to figure out just what's at stake in telling the truth or failing to do so" (Nelson and Nelson 1996:25).

Scenario 4: "I use a wheelchair because I was paralyzed by polio 40 years ago. One of my first trips out of the hospital back then was to a supermarket. I remember I was rolling down an aisle when a kid saw me. He stopped dead in his tracks and pointed. 'Mommy,' he said in a loud voice.....and in a loud voice, 'Mommy, look at the broken man."

Scenario 5: "Lauren M. Cook had been participating in reenactments of famous Civil War battles for two years and she took the hobby seriously. She spent thousands of dollars buying Civil War period clothing. She bound her breasts under her uniform so no one would know she was a woman. She even tried to adopt male mannerisms to aid her disguise. 'I would always squint,' she said. 'Women's eyes are larger than men's, so they really give you away'" (Marcus 2002).

Applied Research

Read *The New York Times, Los Angeles Times*, or some other newspaper with a national reputation for one week. Clip out or print articles about disruptions in the division of labor as defined by Durkheim. Do the articles give insight about how disruptions break down people's ability to connect with one another in meaningful ways through their labor?

Practice Test: Multiple-Choice Questions

1. Focusing on the Congo and its relationship to other countries helps us to
   a. identify the origin of HIV.
   b. understand how AIDS became a global epidemic.
   c. understand why Africa is the "cradle of AIDS."
   d. understand why AIDS is a predominantly homosexual disease in the U.S and throughout the world.

2. "Reality is a social construction" means that
   a. nothing in the world is real.
   b. actions have consequences.
   c. people assign meaning to what is going on around them.
   d. truth is an illusion.

3. When studying social interaction, sociologists strive to establish the content and context of interaction. Content is _______ and context is _______.
   a. cultural frameworks; historical circumstances
   b. historical circumstances; cultural frameworks
   c. historical forces; cultural interpretations
   d. symbolic meanings; cultural frameworks

4. __________ wrote *The Division of Labor in Society*.
   a. Karl Marx
   b. Max Weber
   c. Emile Durkheim
   d. C. Wright Mills

5. Ultimately the Europeans vigorously colonized much of Asia, Africa, and the Pacific in the late nineteenth and early twentieth centuries because of
   a. their superior intellect.
   b. a need to help out the less "civilized" peoples.
   c. a growing demand for resources and low cost (even free) labor.
   d. the human need to explore the unknown.

6. The Mbuti pygmies are a hunting and gathering people who share
   a. an exploitive values system.
   b. a forest-oriented value system.
   c. a capitalist value system.
   d. a Muslim faith.

7. In societies characterized by mechanical solidarity, the ties that bind people together are based primarily on
   a. kinship and religion.
   b. occupation.
   c. agriculture and friendship.
   d. social status and the division of labor.

8. The common conscience of the Mbuti revolves around
   a. the sea.
   b. the forest.
   c. the mines.
   d. nature in general.

9. According to Durkheim, the vulnerability of societies __________ as the division of labor becomes more complex and specialized.
   a. decreases
   b. remains the same
   c. increases
   d. increases but eventually decreases

10. Which one of the following "disruptive events" occurs when workers are so isolated that few people grasp the workings and consequences of the overall enterprise?
    a. job specialization
    b. industrial and commercial crises
    c. inefficient management of worker talents
    d. forced division of labor

11. Historically disease patterns are affected by
    a. economic and social isolation.
    b. the extent to which a society depends on Western medicine.
    c. changes in population density and transportation patterns.
    d. the quality of health care professionals.

12. The American Red Cross, the largest blood supplier in the United States, collects 6.8 million blood donations annually. From a sociological point of view, the amount of blood exchanged represents one indicator of the
    a. role strain associated with giving and receiving blood.
    b. amount of indirect interaction between blood donors and receivers.
    c. role conflict associated with giving and receiving blood.
    d. safety of the blood supply.

13. Sociologists use the word social status to mean
    a. a role.
    b. a rank.
    c. prestige.
    d. a position in a social structure.

14. __________ consists of two or more people interacting and interrelating in specific, expected ways, regardless of their unique personalities.
    a. Role set
    b. Impression management
    c. Social structure
    d. Social behavior

15. When a professor fails to prepare for class
    a. students' rights are violated.
    b. students' obligations are violated.
    c. students do not have to uphold their obligation to study.
    d. social structure collapses.

16. Western medicine is shaped by a profound cultural belief in
    a. the ability of the body to heal on its own.
    b. the ability of technology to solve medical problems.
    c. self-medication.
    d. alternative medicine.

17. As a patient, Shelby has an obligation to follow her doctor's treatment plan. However, the drug prescribed makes her feel very drowsy, preventing her from carrying out her role as a mother. Shelby feels she is not alert enough to properly care for her daughter. Shelby is experiencing
    a. role strain.
    b. status inconsistency.
    c. a role set.
    d. role conflict.

18. When social interaction is viewed as though it were taking place in a theater, that perspective corresponds to the
    a. dramaturgical model.
    b. historical model.
    c. cultural strain model
    d. division of labor model.

19. Which one of the following statements about impression management is <u>false</u>?
    a. People usually are not aware that they are engaged in impression management because they are simply behaving in ways they regard as natural.
    b. Impression management can be a constructive and normal feature of social interaction.
    c. The dark side of impression management emerges when people manipulate the audience in deliberately deceitful and hurtful ways.
    d. If people spoke and behaved as they please, relationships would become more open.

20. The front stage is the area
    a. out of the audience's sight.
    b. where people take care to create and maintain expected images and behavior.
    c. where individuals can "let their hair down."
    d. that people take great care to conceal from the audience.

21. The U.S. supplies about ___________ percent of the world's blood and blood products.
    a. 10
    b. 15
    c. 25
    d. 60

22. Attributing cause to _______ factors functions to reduce uncertainty about the source and spread of disease.
    a. dispositional
    b. situational
    c. backstage
    d. contextual

23. Ramona claims she failed a biology exam because her professor can't explain the subject matter and asks tricky questions. Ramona is attributing her failure to
    a. role strain.
    b. role conflict.
    c. dispositional factors.
    d. situational factors.

24. People usually attribute cause to either dispositional traits or situational factors. Dispositional factors include which one of the following?
    a. bad luck
    b. environmental conditions
    c. personal effort
    d. forces outside an individual's control

25. Throughout history, whenever medical people have lacked knowledge or technology to control a disease, the people in the society have tended to
    a. take humane and responsible steps to combat the disease.
    b. focus on the biological cause of the disease.
    c. act as if a cure will never be found.
    d. hold some group responsible for causing it.

26. When the focus of the cause of AIDS is a specific group (such as homosexuals) and that group's behavior, solutions are framed in terms of
    a. controlling the group.
    b. a global context.
    c. a technological fix.
    d. finding an inexpensive cure.

27. Official definitions of AIDS have all <u>but</u> which of the following consequences?
    a. They affect the way in which the condition is defined.
    b. They influence statistics about who has AIDS.
    c. They affect the content of the physician-patient relationship (e.g., whether the physician asks a patient to be HIV-tested).
    d. They encourage people in low-risk groups to agree to random blood test for HIV-infection.

28. From a sociological point of view, one important question that may provide important clues to understanding AIDS is
    a. Why are young people promiscuous?
    b. Why do some people remain HIV-infected for years and yet have never developed AIDS, while other people develop AIDS shortly after exposure to HIV?
    c. What is it about homosexuals that causes them to engage in high-risk behaviors?
    d. What country is the cradle of AIDS?

Practice Test: True/False Questions

T F 1. In every country or region of the world HIV in the blood supply has preceded or coincided with the first case of AIDS.

T F 2. Some scholars call the Congo the site of Africa's First World War.

T F 3. People <u>enact</u> statuses and <u>occupy</u> roles.

T F 4. It is clear that HIV originated in the Congo and spread to the United States via Europe, Haiti, or Cuba.

T F 5. The United States is the world's largest blood supplier.

T F 6. Neil Postman is the sociologist associated with the dramaturgical model of social interaction.

T F 7. Impression management is always manipulative and deceitful.

T F 8. When evaluating the causes of their own failures people tend to favor situational factors.

T F 9. Approximately 80 percent of AIDS cases in the United States have been traced to more than one mode of exposure.

Internet Resources Related to AIDS/HIV

- **Rethinking AIDS website**
  http://www.virusmyth.com/aids/index.htm
  This site offers many interesting and different perspectives on the cause of AIDS with more than 1200 web pages and over 850 articles. While it is unclear who is responsible for constructing this site, it is clear that its inspiration comes from the *Group for the Scientific Reappraisal of the HIV-AIDS Hypothesis*. The rethinking AIDS website presents evidence supporting a reappraisal. Among other things, it elaborates on the 'HIV as the cause of AIDS" controversy and introduces whistle-blowers—those who question the HIV-AIDS connection. Very thought-provoking and controversial material.

- **Union of Concerned Scientists**
  http://www.ucsusa.org/
  The Union of Concerned Scientists, an independent nonprofit alliance of more than 100,000 concerned citizens and scientists seeks to "build public awareness of global environmental issues" including antibiotic resistance, biodiversity, fuel-efficient vehicles, genetically engineered food, global warming, health and environment, missile defense, nuclear power, and renewable energy.

Internet Resources Related to Democratic Republic of the Congo

- **Democratic Republic of the Congo (Zaire) Page**
  http://www.sas.upenn.edu/African_Studies/Country_Specific/Zaire.html
  The African Studies Center at the University of Pennsylvania posts country-specific pages on sub-Saharan African countries including the Democratic Republic of the Congo (formerly Zaire).

- **Conflicts in Africa; The Democratic Republic of the Congo**
  http://www.globalissues.org/Geopolitics/Africa/DRC.asp
  This website by Anup Shah has links and maps with information on the Democratic Republic of the Congo such as a brief background, the struggle for political power and the international battle over resources.

Democratic Republic of the Congo: Background Notes

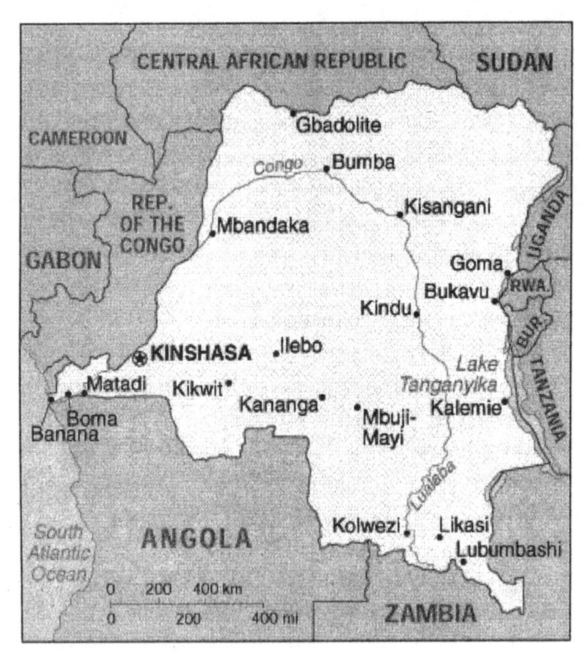

PEOPLE

Nationality: *Noun and adjective*--Congolese.
Population (2004 est.): 58 million.
Annual growth rate (2004 est.): 2.99%.
Ethnic groups: More than 200 African ethnic groups; the Luba, Kongo, and Anamongo are some of the larger groupings of tribes.
Religions (2004 est): Roman Catholic 50%, Protestant 20%, other syncretic sects and traditional beliefs 10%, Kimbanguist 10%, Muslim 10%.
Language: *Official*--French. *National languages*--Lingala, Swahili, Kikongo, Tshiluba.
Education: *Literacy* (2004 est.)--65.5% in French or local language. *Schooling* (2000 est.)--none--41.7%, primary--42.2%, secondary--15.4%, University--0.7%.
Health (2004 est.): *Infant mortality rate*--94.69/1,000 live births. *Life expectancy*--49 yrs.

The population of D.R.C .was estimated at 58 million in 2004. As many as 250 ethnic groups have been distinguished and named. Some of the larger groupings of tribes are the Kongo, Luba, and Anamongo. Although 700 local languages and dialects are spoken, the linguistic variety is bridged both by the use of French and the intermediary languages Kikongo, Tshiluba, Swahili, and Lingala.

About 50% of the Congolese population is Christian, predominantly Roman Catholic. Most of the non-Christians adhere to either traditional religions or syncretic sects. Traditional religions include concepts such as monotheism, animism, vitalism, spirit and ancestor worship, witchcraft, and sorcery and vary widely among ethnic groups; none is formalized. The syncretic sects often merge Christianity with traditional beliefs and rituals. The most popular of these sects, Kimbanguism, was seen as a threat to the colonial regime and was banned by the Belgians. Kimbanguism, officially "the church of Christ on Earth by the prophet Simon Kimbangu," now claims about 3 million members, primarily among the Bakongo tribe of Bas-Congo and Kinshasa. In 1969, it was the first independent African church admitted to the World Council of Churches.

Before independence, education was largely in the hands of religious groups. The primary school system was well developed at independence; however, the secondary school system was limited, and higher education was almost nonexistent in most regions of the country. The principal objective of this system was to train low-level administrators and clerks. Since independence, efforts have been made to increase access to education, and secondary and higher education have been made available to many more Congolese. According to estimates made in 2000, 41.7% of the population has no schooling, 42.2% has primary schooling, 15.4% has secondary schooling, and 0.7% has university schooling. At all levels of education, males greatly outnumber females. The largest state-run universities are the University of Kinshasa, the University of Lubumbashi, and the University of Kisangani. The elite continue to send their children abroad to be educated, primarily in Western Europe.

HISTORY

The area known as the Democratic Republic of the Congo was populated as early as 10,000 years ago and settled in the 7th and 8th centuries A.D. by Bantus from present-day Nigeria. Discovered in 1482 by Portuguese navigator Diego Cao and later explored by English journalist Henry Morton Stanley, the area was officially colonized in 1885 as a personal possession of Belgian King Leopold II as the Congo Free State. In 1907, administration shifted to the Belgian Government, which renamed the country the Belgian Congo. Following a series of riots and unrest, the Belgian Congo was granted its independence on June 30, 1960. Parliamentary elections in 1960 produced Patrice Lumumba as prime minister and Joseph Kasavubu as president of the renamed Democratic Republic of the Congo.

Within the first year of independence, several events destabilized the country: the army mutinied; the governor of Katanga province attempted secession; a UN peacekeeping force was called in to restore order; Prime Minister Lumumba died under mysterious circumstances; and Col. Joseph Désiré Mobutu (later Mobutu Sese Seko) took over the government and ceded it again to President Kasavubu.

Unrest and rebellion plagued the government until 1965, when Lieutenant General Mobutu, by then commander in chief of the national army, again seized control of the country and declared himself president for 5 years. Mobutu quickly centralized power into his own hands and was elected unopposed as president in 1970. Embarking on a campaign of cultural awareness, Mobutu renamed the country the Republic of Zaire and required citizens to adopt African names. Relative peace and stability prevailed until 1977 and 1978 when Katangan rebels, staged in Angola, launched a series of invasions into the Katanga region. The rebels were driven out with the aid of Belgian paratroopers.

During the 1980s, Mobutu continued to enforce his one-party system of rule. Although Mobutu successfully maintained control during this period, opposition parties, most notably the *Union pour la Democratie et le Progres Social (UDPS)*, were active. Mobutu's attempts to quell these groups drew significant international criticism.

As the Cold War came to a close, internal and external pressures on Mobutu increased. In late 1989 and early 1990, Mobutu was weakened by a series of domestic protests, by heightened international criticism of his regime's human rights practices, and by a faltering economy. In April 1990 Mobutu agreed to the principle of a multi-party system with elections and a constitution. As details of a reform package were delayed, soldiers in September 1991 began looting Kinshasa to protest their unpaid wages. Two thousand French and Belgian troops, some of whom were flown in on U.S. Air Force planes, arrived to evacuate the 20,000 endangered foreign nationals in Kinshasa.

In 1992, after previous similar attempts, the long-promised Sovereign National Conference was staged, encompassing more than 2,000 representatives from various political parties. The conference gave itself a legislative mandate and elected Archbishop Laurent Monsengwo as its chairman, along with Etienne Tshisekedi, leader of the UDPS, as prime minister. By the end of the year Mobutu had created a rival government with its own prime minister. The ensuing stalemate produced a compromise merger of the two governments into the High Council of Republic-Parliament of Transition (HCR-PT) in 1994, with

Mobutu as head of state and Kengo Wa Dondo as prime minister. Although presidential and legislative elections were scheduled repeatedly over the next 2 years, they never took place.

By 1996, the war and genocide in neighboring Rwanda had spilled over to Zaire. Rwandan Hutu militia forces (Interahamwe), who fled Rwanda following the ascension of a Tutsi-led government, were using Hutu refugee camps in eastern Zaire as bases for incursions against Rwanda.

In October 1996, Rwandan troops (RPA) entered Zaire, simultaneously with the formation of an armed coalition led by Laurent-Desire Kabila known as the *Alliance des Forces Democratiques pour la Liberation du Congo-Zaire (AFDL)* . With the goal of forcibly ousting Mobutu, the AFDL, supported by Rwanda and Uganda, began a military campaign toward Kinshasa. Following failed peace talks between Mobutu and Kabila in May 1997, Mobutu left the country, and Kabila marched into Kinshasa on May 17, 1997. Kabila declared himself president, consolidated power around himself and the AFDL, and renamed the country the Democratic Republic of Congo (D.R.C.). Kabila's Army Chief and the Secretary General of the AFDL were Rwandan, and RPA units continued to operate tangentially with the D.R.C.'s military, which was renamed the *Forces Armees Congolaises (FAC).*

Over the next year, relations between Kabila and his foreign backers deteriorated. In July 1998, Kabila ordered all foreign troops to leave the D.R.C. Most refused to leave. On August 2, fighting erupted throughout the D.R.C. as Rwandan troops in the D.R.C. "mutinied," and fresh Rwandan and Ugandan troops entered the D.R.C. Two days later, Rwandan troops flew to Bas-Congo, with the intention of marching on Kinshasa, ousting Laurent Kabila, and replacing him with the newly formed Rwandan-backed rebel group called the *Rassemblement Congolais pour la Democratie (RCD)*. The Rwandan campaign was thwarted at the last minute when Angolan, Zimbabwean, and Namibian troops intervened on behalf of the D.R.C. Government. The Rwandans and the RCD withdrew to eastern D.R.C., where they established de facto control over portions of eastern D.R.C. and continued to fight the Congolese Army and its foreign allies.

In February 1999, Uganda backed the formation of a rebel group called the *Mouvement pour la Liberation du Congo (MLC)*, which drew support from among ex-Mobutuists and ex-FAZ soldiers in Equateur province (Mobutu's home province). Together, Uganda and the MLC established control over the northern third of the D.R.C.

At this stage, the D.R.C. was divided de facto into three segments, and the parties controlling each segment had reached military deadlock. In July 1999, a cease-fire was proposed in Lusaka, Zambia, which all parties signed by the end of August. The Lusaka Accord called for a cease-fire, the deployment of a UN peacekeeping operation, MONUC, the withdrawal of foreign troops, and the launching of an "Inter-Congolese Dialogue" to form a transitional government leading to elections. The parties to the Lusaka Accord failed to fully implement its provisions in 1999 and 2000. Laurent Kabila drew increasing international criticism for blocking full deployment of UN troops, hindering progress toward an Inter-Congolese Dialogue, and suppressing internal political activity.

On January 16, 2001, Laurent Kabila was assassinated. He was succeeded by his son, Joseph Kabila. Joseph Kabila reversed many of his father's negative policies; over the next year, MONUC deployed throughout the country, and the Inter-Congolese Dialogue proceeded. By the end of 2002, all Angolan, Namibian, and Zimbabwean troops had withdrawn from the D.R.C. Following D.R.C.-Rwanda talks in South Africa that culminated in the Pretoria Accord in July 2002, Rwandan troops officially withdrew from the D.R.C. in October 2002, although there were continued, unconfirmed reports that Rwandan soldiers and military advisers remained integrated with RCD/G forces in eastern D.R.C. Ugandan troops officially withdrew from the D.R.C. in May 2003.

In October 2001, the Inter-Congolese Dialogue began in Addis Ababa under the auspices of Facilitator Ketumile Masire (former president of Botswana). The initial meetings made little progress and were adjourned. On February 25, 2002, the dialogue was reconvened in South Africa. It included representatives from the government, rebel groups, political opposition, civil society, and Mai-Mai (Congolese local defense militias). The talks ended inconclusively on April 19, 2002, when the government and the MLC brokered an agreement that was signed by the majority of delegates at the dialogue but left out the RCD/G and opposition UDPS party, among others.

This partial agreement was never implemented, and negotiations resumed in South Africa in October 2002. This time, the talks led to an all-inclusive powersharing agreement, which was signed by delegates in Pretoria on December 17, 2002, and formally ratified by all parties on April 2, 2003. Following nominations by each of the various signatory groups, President Kabila on June 30, 2003 issued a decree that formally announced the transitional government lineup. The four vice presidents took the oath of office on July 17, 2003, and most incoming ministers assumed their new functions within days thereafter. This transitional government is slated to remain in place until elections--the first since 1960--are held in 2005.

ECONOMY

GDP (2003): $5.6 billion.
Annual GDP growth rate (2003): 5%.
Per capita GDP (2003): $98.65.
Natural resources: Copper, cobalt, diamonds, gold, other minerals; petroleum; wood; hydroelectric potential.
Agriculture: *Cash crops*--coffee, rubber, palm oil, cotton, cocoa, sugar, tea. *Food crops*--manioc, corn, legumes, plantains, peanuts.
Land use: Agriculture 3%; pasture 7%; forest/woodland 77%; other 13%.
Industry: *Types*--processed and unprocessed minerals; consumer products, including textiles, plastics, footwear, cigarettes, metal products; processed foods and beverages, cement, timber.
Currency: Congolese franc (FC).
Trade: *Exports* (2002)--$1.040 billion. *Products*--diamonds, cobalt, copper, coffee, petroleum. *Partners*—E.U., Japan, South Africa, U.S., China; *Imports* (2002)--$1.216 billion. *Products*--consumer goods (food, textiles), capital equipment, refined petroleum products. *Partners*—E.U., China, South Africa, U.S.
Total external debt (2002): $8.211 billion. (Currently under revision due to HIPC decision point in 2003)

Sparsely populated in relation to its area, the Democratic Republic of the Congo is home to a vast potential of natural resources and mineral wealth. Nevertheless, the D.R.C. is one of the poorest countries in the world, with per capita annual income of about $98 in 2003. This is the result of years of mismanagement, corruption, and war.

In 2001, the Government of the D.R.C. under Joseph Kabila undertook a series of economic reforms to reverse this steep decline. Reforms were monitored by the IMF and included liberalization of petroleum prices and exchange rates and adoption of disciplined fiscal and monetary policies. The reform program reduced inflation from over 500% per year in 2000 to only about 7% at an annual rate in 2003. In June 2002, the World Bank and IMF approved new credits for the D.R.C. for the first time in over a decade. Bilateral donors, whose assistance has been almost entirely dedicated to humanitarian interventions in recent years, also are beginning to fund development projects in the D.R.C. In October 2003, the World Bank launched a multi-sector plan for development and reconstruction. The Paris Club also granted the

D.R.C. Highly Indebted Poor Country status in July 2003. This will help alleviate the D.R.C.'s external sovereign debt burden and potentially free funds for economic development.

Agriculture is the mainstay of the Congolese economy, accounting for 56.3% of GDP in 2002. The main cash crops include coffee, palm oil, rubber, cotton, sugar, tea, and cocoa. Food crops include cassava, plantains, maize, groundnuts, and rice. Industry, especially the mining sector, is underdeveloped relative to its potential in the D.R.C. In 2002, industry accounted for only 18.8% of GDP; with only 3.9% attributed to manufacturing. Services reached 24.9% of GDP. The Congo was the world's fourth-largest producer of industrial diamonds during the 1980s, and diamonds continue to dominate exports, accounting for over half of exports ($642 million) in 2003. The Congo's main copper and cobalt interests are dominated by Gecamines, the state-owned mining giant. Gecamines production has been severely affected by corruption, civil unrest, world market trends, and failure to reinvest.

For decades, corruption and misguided policy have created a dual economy in the D.R.C. Individuals and businesses in the formal sector operated with high costs under arbitrarily enforced laws. As a consequence, the informal sector now dominates the economy. In 2002, with the population of the D.R.C. estimated at 56 million, only 230,000 Congolese working in private enterprise in the formal sector were enrolled in the social security system. Approximately 600,000 Congolese were employed by the government.

In the past year, the Congolese Government has approved a new investment code and a new mining code and has designed a new commercial court. The goal of these initiatives is to attract investment by promising fair and transparent treatment to private business. The World Bank also is supporting efforts to restructure the D.R.C.'s large parastatal sector, including Gecamines, and to rehabilitate the D.R.C.'s neglected infrastructure, including the Inga Dam hydroelectric system.

The outbreak of war in the early days of August 1998 caused a major decline in economic activity. Economic growth, however, resumed in 2002 with a 3% growth rate continuing in 2003 at 5%. The country had been divided de facto into different territories by the war, and commerce between the territories had halted. With the installation of the transitional government in July 2003, the country has been "de jure" reunified, and economic and commercial links have begun to reconnect.

In June 2000, the United Nations established a Panel of Experts on the Illegal Exploitation of Congolese Resources to examine links between the war and economic exploitation. Reports issued by the panel indicate that countries involved in the war in Congo have developed significant economic interests. These interests may complicate efforts by the government to better control its natural resources and to reform the mining sector. A final panel report was issued in October 2003. The Panel of Experts mandate was not renewed.

U.S.-CONGOLESE RELATIONS

Its dominating position in Central Africa makes stability in the D.R.C. an important element of overall stability in the region. The United States supports the transitional government and encourages peace, prosperity, democracy, and respect for human rights in the D.R.C. The United States remains a partner with the D.R.C. and other central African nations in their quest for stability and growth on the continent. From the start of the Congo crisis, the United States has pursued an active diplomatic strategy in support of these objectives. In the long term, the United States seeks to strengthen the process of internal reconciliation and democratization within all the states of the region to promote stable, developing, and democratic nations with which it can work to address security interests on the continent and with which it can develop mutually beneficial economic relations. The United States appointed its current ambassador to the D.R.C. in 2004. The D.R.C. appointed its current ambassador to the United States in 2000. There is no current U.S. direct bilateral aid to the Government of the Congo. USAID's 2004 program in the D.R.C. totals $120 million, which will be used by international and local NGOs for a wide range of relief and developmental activities throughout the country. The Congo has been on the State Department's travel advisory list since 1977.

Source: The U.S. Department of State (2004)
http://www.state.gov

Chapter References

Gallagher, Hugh. 1992. *NPR* "Morning Edition." (July 3).

Marcus, Amy . 2002. "When Janie Came Marchin Home." *The New York Times* (March 23):A17.

Nelson, James Lindemann and Hilde Lindemann. 1996. *Alzheimer's: Answers to Hard Questions for Families*. New York: Doubleday.

Stolberg, Sheryl. 1994. "Doctors' Dilemma." *Los Angeles Times* (April 5):E1+.

Answers

Concept Application

1. Role strain — pg. 154
2. Front stage, Impression management — pg. 159; 157
3. Impression management — pg. 157
4. Dispositional traits — pg. 163
5. Impression management — pg. 157

Multiple-Choice

1.	b	pg. 140
2.	c	pg. 141
3.	a	pg. 142
4.	c	pg. 142
5.	c	pg. 142
6.	b	pg. 144
7.	a	pg. 144
8.	b	pg. 144
9.	c	pg. 147
10.	a	pg. 147
11.	c	pg. 151
12.	b	pg. **
13.	d	pg. 153
14.	c	pg. 153
15.	a	pg. 154
16.	b	pg. 154
17.	d	pg. 154
18.	a	pg. 156
19.	d	pg. 157
20.	b	pg. 159
21.	d	pg. 162
22.	a	pg. 163
23.	d	pg. 163
24.	c	pg. 163
25.	d	pg. 164
26.	a	pg. 164
27.	d	pg. 166
28.	b	pg. 168

** footnote

True/False

1.	T	pg. 141
2.	T	pg. 150
3.	F	pg. 153
4.	F	pg. 164
5.	T	pg. 160
6.	F	pg. 157
7.	F	pg. 157
8.	T	pg. 164
9.	F	pg. 167

Chapter 6

Social Organization
With Emphasis on McDonald's

Study Questions

1. Why is the McDonald's corporation the focus of a chapter on social organization?

worlds largest food service retailer

2. What is an organization? How do sociologists approach the study of organizations?

- a coordinating mechanism created by people to achieve an agreed-upon goal.
  - brings together people, resources, technology ect...
- can be studied apart from the people who form them

3. Define multinational corporation. Give an example.

- enterprises that own/control the production/service facilities in countries other than the one in which their headquaters are located

ex: McDonalds

4. In what ways are multinational corporations engines of progress? In what ways are they engines of destruction? Use the bottled water industry as an example.

Progress - gives jobs
destruction - exploit poor

5. What are externality costs?

- not figured into the price of a product but are eventually paid by consumer through using, creating or desposing of the product

ex: cleaning environment, injured workers

6. What do sociologists mean when they say organizations have two faces or two sides?

(1) capable of efficiently managing people, goods, info worldwide
(2) promote inefficient, destructive actions.

7. What is value-rational thought? Why was Max Weber especially concerned with value-rational action?

- people strive to find most efficient way to achieve a goal.
- doesn't assume better understanding of our surroundings

8. Define rationalization. How do the examples of the cow and potato relate to rationalization? What are the positive and negative outcomes of rationalization?

- action rooted in emotion, tradition are replaced by those grounded in "means to an end"

cow = milk = cheese for pizza hut

9. What is McDonaldization of society?

- principles of fastfood are becoming more dominant in society.

10. Explain the iron cage of rationality.

- irrationalities that rational systems generate

ex divide responsibilities - high turnover rate for employees.

11. What is a bureaucracy? Is McDonald's a bureaucracy? Explain.

- organization that uses most efficient means to achieve a valued goal.
- hierarchical / clear labor division / written rules ...

12. How is studying a bureaucracy as an ideal type useful?

ideal type - simplification of beuracracy in that characteristics exaggerate certain aspects, makes them objects of comparison.

13. Distinguish between formal and informal dimensions of organizations.

formal: official, written guidelines/rules - define goals
informal: generated norms - don't correspond w/ policies

14. What is trained incapacity? Give an example from Shoshana Zuboff's *In the Age of the Smart Machine*, of a work environment that promotes trained incapacity. Contrast that work environment with one that promotes empowering behavior.

– inability to respond to new/unusual circumstances or know when rules are no longer applicable.

15. How do organizations use statistical measures of performance? What are some of the problems that can accompany such measures? Give examples of statistical measures of performance at McDonald's.

– stats on losses, sales, satisfaction, turnover...
– not valid, employees ignore problems to score well

16. What is expert power? How can expert power be problematic?

– superior gives orders to subordinates
– complex

17. Define oligarchy. Why does oligarchy seem to be an inevitable feature of large organizations?

– rule by the few
– democratic participation impossible
– interdependent with technology

18. How did Karl Marx define alienation? What are the four levels of alienation? Give concrete examples of alienation.

– state in which human life is dominated by the forces of human inventions

① process of production (impersonal market)
② from product (specialized role)
③ from family/community (house separate from work)
④ from self (routinize)

Key Concepts

Organizations	pg. 181
Multinational corporations	pg. 185
Rationalization	pg. 188
Externality costs	pg. 188
Bureaucracy	pg. 195
Ideal type	pg. 196
Formal dimension	pg. 196
Informal dimension	pg. 196
Trained incapacity	pg. 197
Professionalization	pg. 200
Oligarchy	pg. 201
Alienation	pg. 202
McDonaldization of society	pg. 194
Efficiency	pg. 194
Quantification and calculation	pg. 194
Predictability	pg. 194
Control	pg. 194
Iron cage of rationality	pg. 195

Concept Application

Consider the concepts listed below. Match one or more of the concepts with each scenario. Explain your choices.

- ✓ Automate
- ✓ Externality costs
- ✓ Informal dimensions of organizations
- ✓ Multinational corporation
- ✓ Value-rational

Scenario 1 "Is IBM Japan an American or a Japanese company? Its work force of 20,000 is Japanese, but its equity holders are American. Even so, over the past decade IBM Japan has provided, on average, three times more tax revenue to the Japanese government than has Fujitsu. What is its nationality? Or what about Honda's operation in Ohio? Or Texas Instruments' memory-chip activities in Japan? Are they 'American' products? If so, what about the cellular phones sold in Tokyo that contain components made in the United States by American workers who are employed by the U.S. division of a Japanese company? Sony has facilities in Dotham, Alabama, from which it sends audio tapes and video tapes to Europe. What is the nationality of these products or the operation that makes them?" (Ohmae 1990:10).

Scenario 2 "A number of employees (5%) respond to perceived injustices by not performing their required tasks. One incident involved a male stockroom worker at a retail store who claimed he was paid less than others in similar positions. After an unsuccessful attempt to discuss the matter with this supervisor, he decided to deal with the conflict in his own way: 'I didn't really want to quit so I goofed off a lot. I didn't do anything unless I was specifically asked to. When working at night I would listen to music for hours and do nothing.... If I was goofing off and saw the manager, I would act as if I was really doing something'" (Tucker 1993:37).

Scenario 3 "Scientifically the atomic bomb was an advance into unknown territory, but militarily it was simply a more cost-effective way of attaining a goal that was already a central part of strategy: a means of producing the results achieved at Hamburg and Dresden cheaply and reliably every time the weapon was used [for example, a quarter million bombs were used to destroy the city of Dresden]. (Even at the time, the $2 billion cost of the Manhattan Project was dwarfed by the cost of trying to destroy cities the hard way, using conventional bombs)" (Dyer 1985:96).

Scenario 4 "The same kind of computer technology that enables employers to keep track of workers' backgrounds also makes it possible for them to quantify and monitor work performance. Anyone who works on a video display terminal, electronic telephone console, or other computer-based equipment, including laser scanner cash registers, is subject to constant monitoring.

Although the stated aim of monitoring workers is to improve productivity and service, the effect can be to turn checkstands into pressure cookers.

'Computers are wonderful for many things,' say Beverly Crownover, president of Local 1532 of United Food and Commercial Workers in Santa Rosa, California. 'But when they're used to monitor how many items a cashier scans per minute, it's like a whip. There's incredible pressure on workers'" (*UFCW Action* 1993:135).

Scenario 5 The cost of stress to the American workplace has been estimated at between $150 billion and $180 billion a year. Stress-related illness accounts for millions of lost working days each year, and the number is rising. One study found that in 1980 no occupational disease claims were related to stress; in 1990, 10 percent of them were. A 1993 study by Commerce Clearing House reports that unscheduled absences can cost U.S. employers more than $500 employee per year. Experts believe that stress accounts for 12 percent of all workers' compensation claims (Wright and Smye 1996:7).

Applied Research

Select one of the world's top 25 largest global corporations listed in your textbook. Research how the corporation you selected reaches around the globe through joint ventures, supply deals for parts and components, assembly operations, and marketing and distribution offices.

Practice Test: Multiple-Choice Questions

1. The McDonald's corporation has franchises in approximately ______ countries.
   a. 55
   b. 120
   c. 220
   d. 300

2. Which of the following is the <u>best</u> example of an organization?
   a. the class of '95
   b. shoppers in a mall
   c. Wal-Mart
   d. the country of India

3. Outside of the United States, which country is home to the greatest number of McDonalds?
   a. Korea
   b. Japan
   c. Australia
   d. South Africa

4. The world's largest global corporation in 2004 was
   a. General Electric.
   b. IBM.
   c. Exxon/Mobil.
   d. Wal-Mart.

5. Which one of the following is <u>not</u> one of the strategies McDonald's used to become a global giant?
   a. lower production costs.
   b. create new products people need to buy.
   c. create new markets.
   d. accepting cash only.

6. Order takers at a fast food restaurant ask every customer whether he or she would like to "supersize" their drink or meal. This strategy is an example of
   a. lowering production costs.
   b. creating new products.
   c. creating new markets.
   d. suggestive selling.

7. Max Weber's ideas about organizations are built on an understanding of
   a. rationalization.
   b. disenchantment.
   c. coordinating mechanisms.
   d. instrumental action.

8. Weber defined rationalization as a process whereby thought and action rooted in emotion are replaced by
   a. value-rational thought and action.
   b. mysterious forces.
   c. tradition.
   d. instrumental action.

9. Weber made several important qualifications regarding value-rational thought and action. Which of the following is one of them?
   a. Rationalization refers to the way people actually think.
   b. Rationalization refers to the ways in which daily life is organized to accommodate large numbers of people.
   c. Organizations cannot accommodate large numbers of people.
   d. On a personal level, people have no experience with rationalization.

10. According to the Pizza Hut website, the company needs a herd of 250,000 dairy cows producing at full capacity 365 days a year to fill the company's demand for cheese toppings. This view of cows reflects __________ action.
   a. instrumental
   b. value-rational
   c. traditional
   d. effective

11. A perfectly rational organization is a/an
   a. informal organization.
   b. bureaucracy.
   c. formal organization.
   d. oligarchy.

12. According to standard operating procedures, every customer at McDonald's is greeted with the words "Welcome to McDonald's. May I take your order?" The practice corresponds with which characteristic of a bureaucracy?
    a. a clean-cut division of labor
    b. positions filled on the basis of qualification
    c. personnel treat "clients" as cases
    d. authority belongs to the position

13. The ______ dimension of organizations consists of the official, written guidelines, rules, regulations, and policies.
    a. formal
    b. informal
    c. scientific
    d. traditional

14. McDonald's 600-page operations and training manual (that specifies everything from where sauces should be placed on buns to how thick pickle slices should be) would be of interest to sociologists studying
    a. oligarchy.
    b. expert authority.
    c. the informal dimensions of organizations.
    d. the formal dimensions of organizations.

15. In her book *In the Age of the Smart Machine*, Zuboff distinguished between work environments that promote trained incapacity and those that promote
    a. freedom of expression.
    b. empowering behavior.
    c. a clear division of labor.
    d. flextime.

16. A worker says, "Sometimes, I am amazed when I realize that we stare at the computer screen even when it has gone down." This comment suggests that in that organization, computers are used as
    a. an automating tool.
    b. an informating tool.
    c. a coordinating mechanism.
    d. a technological resource.

17. Workers at a manufacturing plant had no knowledge of the production process as a whole or the rationale behind any of the rules and procedures. Most workers reported that they were trained to master certain steps but not to handle the machinery under all conditions. Such a situation is an example of
    a. trained incapacity.
    b. expert authority.
    c. oligarchy.
    d. rationalization.

18. Encouraging employees to concentrate on achieving good scores and to ignore problems generated by their drive to score well illustrates the problem with
    a. statistical measures of performance.
    b. trained incapacity.
    c. oligarchy.
    d. expert power.

19. June works as a cashier. Her productivity is judged according to the number of items passed over a scanner per hour. She is being rated according to
    a. statistical measures of performance.
    b. trained incapacity.
    c. informal policies.
    d. an oligarchy.

20. University of California researchers created a technology to produce growth hormones to treat dwarfism. Corporations market the product as a solution for people dissatisfied with their height. This situation speaks to the problem of
    a. trained incapacity.
    b. statistical measures of performance.
    c. informal dimension of organization.
    d. expert knowledge and responsibility.

21. Oligarchy is
    a. the concentration of decision-making power in the hands of a few people.
    b. expert power.
    c. a body of persons organized and classified according to rank.
    d. government run by the clergy.

22. The danger of oligarchy is that those who make decisions may
    a. run the organization as a bureaucracy.
    b. not have the necessary background to understand the full implications of their decisions.
    c. rely on informal mechanisms to get things done.
    d. suffer from disenchantment of the world.

23. Karl Marx believed that increased control over nature is accompanied by
    a. disenchantment.
    b. anomie.
    c. alienation.
    d. trained incapacity.

24. Which one of the following circumstances is <u>not</u> an example of alienation?
    a. Families are forced to move where work is available.
    b. Workers are treated as economic components rather than as active, creative social beings.
    c. Employers use technology as an informating tool.
    d. No person can claim a product as the unique result of his or her labor.

25. Carmen works between 6pm and 2am, 6 days a week. Thus, she is not home for dinner and is still sleeping when her husband leaves for work and her kids leave for school. Marx would argue that such a situation leaves Carmen alienated from
    a. the self.
    b. the family.
    c. the product.
    d. the process of production.

Practice Test: True/False Questions

T F 1. McDonald's is listed among the *Fortune 500*-top U.S.-based corporations.

T F 2. In this chapter, we emphasize McDonald's because it is the worst fast food organization in the United States.

T F 3. Organizations have a life that extends beyond the people who make them up.

T F 4. Multinational corporations are headquartered disproportionately in the United States, Japan, and Western Europe.

T F 5. The ideal is a standard against which real cases can be measured.

T F 6. From a strictly bureaucratic point of view, emotion interferes with the efficient delivery of goods and services.

T F 7. The iron cage of rationality is a term used to describe the informal dimensions of organizations.

Internet Resources Related to the Multinational Corporation

- **The Global Web100**
  http://metamoney.com/globalListIndex.html
  Global Web100 offers a list of links to the 100 largest non-U.S. based global corporations with webpages on the internet. See http://metamoney.com/usListIndex.html for the 100 largest U.S.-based corporations.

- **Multinational Monitor**
  http://www.essential.org/monitor/monitor.html
  Multinational Monitor is "a monthly magazine devoted primarily to examining the activities of multinational companies." Each issue features a special topic. Examples include "Medicine and Market" "The Case Against GE," "Corporations and the U.S. Poor," and "The Ten Worst Corporations of 2003."

Internet Resources Related to McDonalds and Other Such Service Establishments

- **McDonalds.com**
  http://www.mcdonalds.com
  The official website of the McDonald's corporation offers its Annual Report, Investor Fact Sheet, and other shareholder information. There is also information posted about McDonald's menus including ingredient lists.

- **McSpotlight**
  http://www.mcspotlight.org
  The McSpotlight website is "dedicated to compiling and disseminating factual, accurate, up-to-date information - and encouraging debate - about the workers, policies, and practices of the McDonald's Corporation and all they stand for."

Chapter References

Anderson, Patricia. 1991. *Affairs in Order: A Complete Resource Guide to Death and Dying*. New York: Macmillan.

Dyer, Gwynne. 1985. *War*. New York: Crown.

Ohmae, Kenichi. 1990. *The Borderless World: Power and Strategy in the Interlinked Economy*. New York: Harper Business.

Tucker, James. 1993. "Everyday Forms of Employee Resistance." *Sociological Forum* 8(1):25-45.

UFCW Action. 1993. "The Boss Is Watching." *Utne Reader* (May/June):134-35.

Wright, Lesley and Marti Smye. 1996. *Corporate Abuse*. New York: Macmillan.

Answers

Concept Application

1. Multinational corporation pg. 182
2. Informal dimension of organizations pg. 196
3. Value-rational pg. 192
4. Automate pg. 198
5. Externality costs pg. 188

Multiple-Choice

1.	b	pg. 177	14.	d	pg. 196
2.	c	pg. 181	15.	b	pg. 198
3.	b	pg. 179	16.	a	pg. 198
4.	d	pg. 183	17.	a	pg. 198
5.	d	pg. 184	18.	a	pg. 198
6.	d	pg. 184	19.	a	pg. 198
7.	a	pg. 188	20.	d	pg. 200
8.	a	pg. 188	21.	a	pg. 201
9.	b	pg. 192	22.	b	pg. 202
10.	b	pg. 192	23.	c	pg. 202
11.	b	pg. 195	24.	c	pg. 203
12.	c	pg. 196	25.	b	pg. 203
13.	a	pg. 196			

True/False

1.	T	pg. 201
2.	F	pg. 180
3.	T	pg. 181
4.	T	pg. 182
5.	T	pg. 196
6.	T	pg. 196
7.	F	pg. 185

Chapter 7

Deviance, Conformity, and Social Control
With Emphasis on the People's Republic of China

Study Questions

1. Why focus on China in conjunction with concepts of deviance, conformity, and social control?

cultural revolution

2. What is deviance? How is it related to conformity and social control?

- any behavior that is socially challenged/condemned because it departs from norms/expectations
- follow maintain standards
- methods to persuade members to comply/not deviate.

3. Is it possible to generate a list of deviant behavior? Why or why not?

no – everything is qualified as deviant under some circumstances

4. What is the unique contribution of sociology to the study of deviance? What two fundamental assumptions guide the sociological perspective on deviance?

emphasize the content under which deviance occurs

5. Describe how ideological commitment is connected to issues of deviance, conformity, and mechanisms of social control in China.

6. Briefly describe the Cultural Revolution. Why was making money a punishable crime during the Cultural Revolution but not after Mao's death?

7. Distinguish between folkways and mores. Give examples of each concept.

- norms apply to routine matters / details of life
  ex how to eat
- norms essential to well being, enforced
  ex breaking the law

8. What important cultural lessons are incorporated into the daily activities of Chinese and American preschoolers?

9. What are the major mechanisms of social control? Why do all societies have such mechanisms in place?

- sanctions - reactions of approval / disproval
- informal - unofficial formal - definite rules - enforced
- censorship - prevent info from reaching audience
- surveillance - monitoring peoples actions.

10. Briefly explain the following statement: "In China social control is everywhere and involves everyone."

11. According to Durkheim, why is crime a "normal" and necessary phenomenon?

- acts that are minor to one may be offensive to another

12. Labeling theorists believe that rules are socially constructed and that members of social groups do not enforce them in uniform or consistent ways. Explain.

13. What are the implications of the categories "secret deviant" and "falsely accused" for the study of deviance?

- people who broke rules but go unnoticed
- people who didn't break law but treated like they did

14. Under which circumstances are people most likely to be falsely accused of a crime?

- eyewitness error
- police coverup

15. What are witch hunts? Why do they occur? Give an example.

- identify behavior believed to be undermining a group/country

16. What is white-collar crime? Why are white-collar criminals less likely to be caught than "common" criminals?

- committed by persons of high social status

17. In Milgram's classic experiment *Obedience to Authority*, why did a significant number of volunteers come to accept an authority's definition of deviance and administer shocks although they caused obvious harm to confederates?

18. Compare and contrast Chinese and American conceptions of guilt or innocence and of the role of lawyers.

19. Who are claims makers? What factors determine a claim maker's success?

- people who articulate/promote claims and tend to gain in some way if audience accepts it as the truth

20. Describe the constructionist approach to analyzing claims makers and claimsmaking activities.

- focuses on process by which specific groups / activities / conditions / artifacts become defined as a problem

21. What is structural strain? What are the sources of structural strain in the United States?

- situation in which:
  - valued goals have unclear limits
  - people are unsure means will allow to achieve goals
  - legit opportunities remain closed to portion of pop

22. What are the responses to structural strain?

- conformity - follow group standards
- innovation - accept but reject means to obtain goal
- ritualism - reject goals - achieve to means
- retreatism - reject both
- rebellion full rejection of both - introduce new

23. Identify one source of structural strain in China. Use Merton's typology of responses to consider how people in China response to this strain.

24. Summarize the major assumptions underlying the theory of differential association. How does this assumption relate to mechanisms of social control in China?

25. What larger historical and geographical factors explain the differences between the Chinese and the American systems of social control?

26. Why is deviance a complex concept?

Key Concepts

Deviance	pg. 210
Conformity	pg. 210
Social control	pg. 210
Folkways	pg. 213
Mores	pg. 213
Sanctions	pg. 219
Positive	pg. 219
Negative	pg. 219
Formal	pg. 219
Informal	pg. 219
Censors	pg. 219
Censorship	pg. 219
Surveillance	pg. 221
Labeling theory	
Conformity	pg. 210
Conformists	pg. 223
Pure deviant	pg. 223
Master status of deviant	pg. 223
Secret deviant	pg. 223
Falsely accused	pg. 224
Witch-hunt	pg. 224
Crime	pg. 219
White-collar crime	pg. 227
Corporate crime	pg. 227
Confederate	pg. 229

Constructionist approach	pg. 230
Claims makers	pg. 231
Structural strain	pg. 237
Conformity	pg. 238
Innovation	pg. 238
Ritualism	pg. 238
Retreatism	pg. 238
Rebellion	pg. 239
Differential association	pg. 242
Deviant subcultures	pg. 242

Concept Application

Consider the concepts listed below. Match one or more of the concepts with each scenario. Explain your choices.

- ✓ Claims makers
- ✓ Falsely accused
- ✓ Mores
- ✓ Secret deviants
- ✓ White-collar crimes

Scenario 1 "The Tobacco Institute was founded in 1958, even before the first Surgeon General's report on the health risks of smoking, to represent the interests of tobacco companies to lawmakers. Once financed by a dozen companies, it now works for only five—Philip Morris, R. J. Reynolds, Lorillard, Liggett and American Brands—but its twofold mission remains the same: to persuade Federal, state and local authorities to lay off and to sell the virtues of the industry to the American public. A staff of lobbyists handles the first task and Ms. Dawson, at 32, the second. The job description is fairly typical for a trade organization—to develop and articulate the industry position on any given issue, then make sure the message reaches the public. But this is no typical industry" (Janofsky 1994:8F).

Scenario 2 "Boesky told the government about his insider trading activities, not only with me, but with at least one other well-known investment banker. Beyond that, he detailed various schemes, concocted with those in the highest circles of power, to circumvent SEC regulations and tax laws. Said Carroll, 'He has played fast and loose with the rules that govern our markets, with the effect of manipulating the outcome of financial transactions measured in the hundreds of millions of dollars'" (Levine and Hoffer 1991:346).

Scenario 3 "The small-time criminals are everywhere. Maybe they're sneaking into more than one theater in the local cineplex. Or grabbing a handful of yogurt peanuts from the grocery store bin and eating all the evidence before getting to the check-out stand. Or making personal long-distance calls from work" (Tomashoff 1993:E1).

Scenario 4 "Death sentences for people who later prove to be innocent are less unusual than is commonly supposed. Just in the last five months, four once-condemned prisoners have been released after spending years on death row. Two of them, in Alabama and Texas, turned out to have been convicted on fabricated evidence and perjured testimony; the third, in Texas, was convicted because of withheld evidence; the fourth, in Maryland, was exonerated by DNA analysis, a technology that was unavailable at the time of his trial" (*The New Yorker* 1993:14).

Scenario 5 "Can a court force an unwilling person to give up part of his or her body (e.g., bone marrow, a kidney) to a relative who needs that body part to survive? That was the question recently brought before the court of common pleas in Allegheny County, PA. The common law has consistently held that one human being is under no legal obligation to give aid or take action to save another human being or to rescue one. The court said that such a rule, although revolting in a moral sense, is founded upon the very essence of a free society, and while other societies may view things differently, our society has as its first principle respect for the individual—and society and government exist to protect that individual from being invaded and hurt by another" (Chayet 1983).

Applied Research

The internet has more than 20,000 crime-stopper sites. The Peel Regional Police website defines Crime Stoppers as a "nonprofit organization which rallies the community, the news media and the police in a collective campaign against crime. Crime Stoppers' mandate is to fight crime." Use one of the internet search engines (i.e., Infoseek, Yahoo, Excite) to find and review a small sample of crime stopper websites. After viewing these websites, consider how Durkheim would react to and explain their existence.

Practice Test: Multiple-Choice Questions

1. The People's Republic of China represents an interesting case for studying issues of deviance. conformity, and social control because
   a. since July 1, 1997 Hong Kong has imposed its system of social control on China.
   b. many of the behaviors that constituted deviance during the Cultural Revolution no longer apply today.
   c. China is attempting to model its system of social control after the U.S. system.
   d. China has a model system of social control.

2. The only characteristic common to all forms of deviance is the fact that
   a. they invoke formal sanctions.
   b. everyone in the society is offended by the behavior.
   c. the behaviors are considered deviant across time and place.
   d. some social audience regards the act or appearance as deviant.

3. The remarkable economic changes that have taken place in China since 1976
   a. suggest that China is moving toward establishing a capitalist economy.
   b. show the problem with Soviet-style centrally-planned economies.
   c. reflect a more dynamic and flexible economy with market elements but still under Communist control.
   d. have been greatly exaggerated by the Western media.

4. China has run a trade surplus with the United States since
   a. 1977
   b. 1985
   c. 1995
   d. 2004

5. When guests depart in the People's Republic of China, the Chinese host walks them out to their vehicle and then stands and waves until the visitors are out of sight. This behavior is an example of a
   a. folkway.
   b. more.
   c. sanction.
   d. mechanism of social control.

6. In China, it was once unthinkable that an individual could marry, have a baby, or obtain housing without first obtaining approval from the Communist Party-controlled work unit or neighborhood committee. Such norms can be classified as
   a. folkways.
   b. mores.
   c. mechanisms of social control.
   d. sanctions.

7. In comparison to American preschools, Chinese preschools are
   a. loosely structured.
   b. social-minded.
   c. self-directed.
   d. materialistic.

8. A teacher asks her class, "Who would like to paint?" and "Who would like to work on a puzzle?" This exchange is most typical of preschools in
   a. China.
   b. Taiwan.
   c. Japan.
   d. the United States.

9. Jeremy wore barrettes to nursery school. One day a boy repeatedly told Jeremy that "only girls wear barrettes." The incident shows how __________ work as mechanisms of social control.
   a. formal positive sanctions
   b. informal positive sanctions
   c. formal negative sanctions
   d. informal negative sanctions

10. __________________ is a method employed to prevent information from reaching some audience.
    a. Censorship
    b. Surveillance
    c. Negative sanctions
    d. Positive sanctions

11. Kai Erikson wrote "the critical variable in the study of deviance, then, is the social audience rather than the individual actor, since the social audience decides whether or not a behavior is deviant." This statement best corresponds with which theory of deviance?
    a. functionalist
    b. labeling theory
    c. differential association
    d. structural strain theory

12. Which one of the 50 states has the highest incarceration rates?
    a. Minnesota
    b. Texas
    c. Louisiana
    d. California

13. When labeling theorists study deviance they focus <u>least</u> on
    a. the context.
    b. the rule-makers.
    c. the rule-enforcers.
    d. the rule that is broken.

14. Labeling theorists maintain that whether an act is deviant depends on all <u>but</u> which one of the following?
    a. whether people notice it
    b. whether people react to it as a violation
    c. whether the deviant act was actually committed by the person accused
    d. whether sanctions are applied

15. Howard Becker wrote "no one really knows how much this phenomenon exists but the amount is very sizable, much more than we are apt to think." Becker was writing about
    a. secret deviants.
    b. innovators.
    c. conformists.
    d. witch-hunts.

16. __________ are likely to be accused of a crime when the well-being of a country or group is threatened.
    a. Pure deviants
    b. The falsely accused
    c. Secret deviants
    d. Conformists

17. Janice Tucker, who works for Lab, Inc., pleaded guilty to submitting false lab analyses of contamination at "clean up" sites. Lab, Inc. charged companies $6,000 for these analyses. Tucker submitted false results because she wanted return business. She committed a/an
    a. corporate crime.
    b. act of retreatism.
    c. white-collar crime.
    d. act of rebellion.

18. China's legal system revolves around
    a. impression management.
    b. charging high fees.
    c. crime control.
    d. due process.

19 In Stanley Milgram's classic experiment, *Obedience to Authority*, Milgram found that obedience was founded on
    a. the firm command of a person with a status that gave minimal authority over a subject recruited to participate in the study.
    b. the subject's fear of being punished physically if he or she disobeyed.
    c. the subject's dislike of the learner's physical characteristics.
    d. the subject's firm belief that learning is enhanced when failure is punished.

20. An article titled "Dark Figures and Child Victims: Statistical Claims about Missing Children" is likely to have been written from a _____ perspective.
    a. constructionist
    b. structural strain
    c. differential association
    d. conformist

21. The Chinese government issued the report "The Human Rights Record of the United States" because
    a. it believes that the United States is a model with regard to human rights.
    b. the U.S. asked for an independent evaluation of its human rights record.
    c. it wanted to show that its human rights record is better than the U.S.
    d. the U.S. issues Country Reports on Human Rights Practices each year on 190 countries but does not critique its own record.

22. Which of the following phrases best summarizes "innovation"?
    a. Win by the rules of the game.
    b. I don't like the game or the rules.
    c. Change the rules to win the game.
    d. Follow the rules even if you don't win.

23. The __________ is one major factor supporting the rigid system of social control in China.
    a. for-profit prison industry in China
    b. rebellious nature of the Chinese people
    c. size of the Chinese population
    d. apathetic nature of the Chinese people

24. The population of China is approximately
    a. 500 million.
    b. 750 million.
    c. 1.3 billion.
    d. 2 billion.

25. The so-called never-imprisoned population includes
    a. pure deviants and conformists.
    b. the falsely accused and secret deviants.
    c. secret deviants and conformists.
    d. conformists and pure deviants.

True/False Questions

T F 1. The U.S. State Department classifies China's human rights record as "fair."

T F 2. Mao used the Cultural Revolution as an attempt to eliminate anyone who opposed his policies.

T F 3. Censorship and surveillance are methods of social control.

T F 4. According to Emile Durkheim, deviance will be present even in a community of saints.

T F 5. China's rate of incarceration (166 per 100,000 population) is the highest in the world.

T F 6. Secret deviants have broken no rules but are treated as if they have.

T F 7. Approximately 50 percent of people sentenced to U.S. federal prisons are classified as drug offenders.

T F 8. About 90 percent of crime victims in the U.S. report the crime to police.

T F 9. One source of structural strain in China relates to the number of legitimate opportunities open to married couples to have children.

T F 10. No regime in China has ever relinquished its power without first resorting to blood shed.

T F 11. Governor George Ryan ordered a moratorium on executions in his state after discovering 13 death row inmates were found innocent by DNA evidence.

Internet Resources Related to Deviance, Conformity, and Social Control

- **Bureau of Justice Statistics**
  http://www.ojp.usdoj.gov/bjs
  The Bureau of Justice Statistics "collects, analyzes, publishes, and disseminates information on crime, criminal offenders, victims of crime, and the operation of justice systems at all levels of government." According to the Bureau the data it collects is pivotal to "combating crime and ensuring that justice is both efficient and evenhanded."

- **The Innocence Project**
  http://www.innocenceproject.org/
  A non-profit organization that has exonerated 154 convicted prisoners on death row since 1992. The organization fights for "poor and forgotten" clients by using biological DNA evidence to prove innocence.

Internet Resources Related to People's Republic of China

- **China News Digest – Global**
  http://www.cnd.org/CND-Global/
  China News Digest is published three times a week. "The contents are news and information about China [and issues related to China and that affect China] that are of general interest to readers all over the world."

- **The China Daily**
  http://www.chinadaily.com.cn/english/home/index.html
  This daily China news site serves as "an online bridge between China and the rest of the world" featuring "news, information, services and education to four million viewers a day."

People's Republic of China: Background Notes

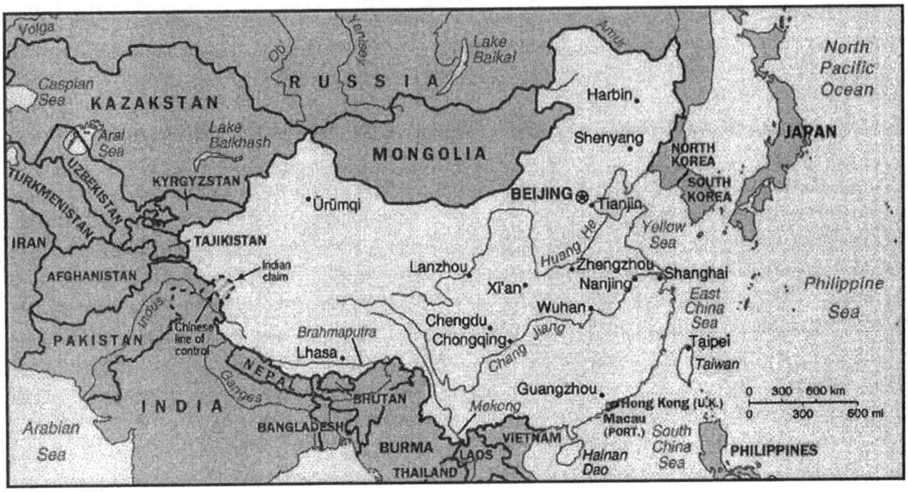

PEOPLE

Nationality: *Noun and adjective*--Chinese (singular and plural).
Population (2003 est.): 1.3 billion.
Population growth rate (2003 est.): 0.6%.
Health (2003 est.): *Infant mortality rate*--25.26/1,000. *Life expectancy*--72.22 years (overall); 70.33 years for males, 74.28 years for females.
Ethnic groups: *Han Chinese*--91.9%; Zhuang, Manchu, Hui, Miao, Uygur, Yi, Mongolian, Tibetan, Buyi, Korean, and *other nationalities*--8.1%.
Religions: Officially atheist; Taoism, Buddhism, Islam, Christianity.
Language: Mandarin (Putonghua), plus many local dialects.
Education: *Years compulsory*--9. Literacy--86%.
Work force (2001 est., 711 million): *Agriculture and forestry*--50%; *industry and commerce*--23%; *other*--27%.

Ethnic Groups

The largest ethnic group is the Han Chinese, who constitute about 91.9% of the total population. The remaining 8.1% are Zhuang (16 million), Manchu (10 million), Hui (9 million), Miao (8 million), Uygur (7 million), Yi (7 million), Mongolian (5 million), Tibetan (5 million), Buyi (3 million), Korean (2 million), and other ethnic minorities.

Language

There are seven major Chinese dialects and many subdialects. Mandarin (or Putonghua), the predominant dialect, is spoken by over 70% of the population. It is taught in all schools and is the medium of government. About two-thirds of the Han ethnic group are native speakers of Mandarin; the rest, concentrated in southwest and southeast China, speak one of the six other major Chinese dialects. Non-Chinese languages spoken widely by ethnic minorities include Mongolian, Tibetan, Uygur and other Turkic languages (in Xinjiang), and Korean (in the northeast).

The Pinyin System of Romanization

On January 1, 1979, the Chinese Government officially adopted the pinyin system for spelling Chinese names and places in Roman letters. A system of Romanization invented by the Chinese, pinyin has long been widely used in China on street and commercial signs as well as in elementary Chinese textbooks as an aid in learning Chinese characters. Variations of pinyin also are used as the written forms of several minority languages.

Pinyin has now replaced other conventional spellings in China's English-language publications. The U.S. Government also has adopted the pinyin system for all names and places in China. For example, the capital of China is now spelled "Beijing" rather than "Peking."

Religion

Religion plays a significant part in the life of many Chinese. Buddhism is most widely practiced, with an estimated 100 million adherents. Traditional Taoism also is practiced. Official figures indicate there are 20 million Muslims, 5 million Catholics, and 15 million Protestants; unofficial estimates are much higher.

While the Chinese constitution affirms religious toleration, the Chinese Government places restrictions on religious practice outside officially recognized organizations. Only two Christian organizations--a Catholic church without official ties to Rome and the "Three-Self-Patriotic" Protestant church--are sanctioned by the Chinese Government. Unauthorized churches have sprung up in many parts of the country and unofficial religious practice is flourishing. In some regions authorities have tried to control activities of these unregistered churches. In other regions, registered and unregistered groups are treated similarly by authorities and congregations worship in both types of churches. Most Chinese Catholic bishops are recognized by the Pope, and official priests have Vatican approval to administer all the sacraments.

Population Policy

With a population officially just under 1.3 billion and an estimated growth rate of about 0.6%, China is very concerned about its population growth and has attempted with mixed results to implement a strict family planning policy. The government's goal is one child per urban family, and two children per rural family, with guidelines looser for ethnic minorities with small populations. Enforcement varies widely, and relies upon "social compensation fees" for extra children as a means of keeping families small. Official government policy opposes forced abortion or sterilization, but in some localities there are instances of forced abortion. The government's goal is to stabilize the population in the first half of the 21st century, and current projections are that the population will peak at around 1.6 billion by 2050.

HISTORY

Dynastic Period

China is the oldest continuous major world civilization, with records dating back about 3,500 years. Successive dynasties developed a system of bureaucratic control that gave the agrarian-based Chinese an advantage over neighboring nomadic and hill cultures. Chinese civilization was further strengthened by the development of a Confucian state ideology and a common written language that bridged the gaps among the country's many local languages and dialects. Whenever China was conquered by nomadic tribes, as it was by the Mongols in the 13th century, the conquerors sooner or later adopted the ways of the "higher" Chinese civilization and staffed the bureaucracy with Chinese.

The last dynasty was established in 1644, when the Manchus overthrew the native Ming dynasty and established the Qing (Ch'ing) dynasty with Beijing as its capital. At great expense in blood and treasure, the Manchus over the next half century gained control of many border areas, including Xinjiang, Yunnan, Tibet, Mongolia, and Taiwan. The success of the early Qing period was based on the combination of Manchu martial prowess and traditional Chinese bureaucratic skills.

During the 19th century, Qing control weakened, and prosperity diminished. China suffered massive social strife, economic stagnation, explosive population growth, and Western penetration and influence. The Taiping and Nian rebellions, along with a Russian-supported Muslim separatist movement in Xinjiang, drained Chinese resources and almost toppled the dynasty. Britain's desire to continue its illegal opium trade with China collided with imperial edicts prohibiting the addictive drug, and the First Opium War erupted in 1840. China lost the war; subsequently, Britain and other Western powers, including the United States, forcibly occupied "concessions" and gained special commercial privileges. Hong Kong was ceded to Britain in 1842 under the Treaty of Nanking, and in 1898, when the Opium Wars finally ended, Britain executed a 99-year lease of the New Territories, significantly expanding the size of the Hong Kong colony.

As time went on, the Western powers, wielding superior military technology, gained more economic and political privileges. Reformist Chinese officials argued for the adoption of Western technology to strengthen the dynasty and counter Western advances, but the Qing court played down both the Western threat and the benefits of Western technology.

Early 20th Century China

Frustrated by the Qing court's resistance to reform, young officials, military officers, and students--inspired by the revolutionary ideas of Sun Yat-sen–began to advocate the overthrow of the Qing dynasty and creation of a republic. A revolutionary military uprising on October 10, 1911, led to the abdication of the last Qing monarch. As part of a compromise to overthrow the dynasty without a civil war, the revolutionaries and reformers allowed high Qing officials to retain prominent positions in the new republic. One of these figures, Gen. Yuan Shikai, was chosen as the republic's first president. Before his death in 1916, Yuan unsuccessfully attempted to name himself emperor. His death left the republican government all but shattered, ushering in the era of the "warlords" during which China was ruled and ravaged by shifting coalitions of competing provincial military leaders.

In the 1920s, Sun Yat-sen established a revolutionary base in south China and set out to unite the fragmented nation. With Soviet assistance, he organized the Kuomintang (KMT or "Chinese Nationalist People's Party"), and entered into an alliance with the fledgling Chinese Communist Party (CCP). After Sun's death in 1925, one of his protégés, Chiang Kai-shek, seized control of the KMT and succeeded in bringing most of south and central China under its rule. In 1927, Chiang turned on the CCP and executed many of its leaders. The remnants fled into the mountains of eastern China. In 1934, driven out of their mountain bases, the CCP's forces embarked on a "Long March" across some of China's most desolate terrain to the northwestern province of Shaanxi, where they established a guerrilla base at Yan'an.

During the "Long March," the communists reorganized under a new leader, Mao Zedong (Mao Tse-tung). The bitter struggle between the KMT and the CCP continued openly or clandestinely through the 14-year long Japanese invasion (1931-45), even though the two parties nominally formed a united front to oppose the Japanese invaders in 1937. The war between the two parties resumed after the Japanese defeat in 1945. By 1949, the CCP occupied most of the country.

Chiang Kai-shek fled with the remnants of his KMT government and military forces to Taiwan, where he proclaimed Taipei to be China's "provisional capital" and vowed to reconquer the Chinese mainland. The KMT authorities on Taiwan still call themselves the "Republic of China."

The People's Republic of China

In Beijing, on October 1, 1949, Mao Zedong proclaimed the founding of the People's Republic of China (P.R.C.). The new government assumed control of a people exhausted by two generations of war and social conflict, and an economy ravaged by high inflation and disrupted transportation links. A new political and economic order modeled on the Soviet example was quickly installed.

In the early 1950s, China undertook a massive economic and social reconstruction program. The new leaders gained popular support by curbing inflation, restoring the economy, and rebuilding many war-damaged industrial plants. The CCP's authority reached into almost every aspect of Chinese life. Party control was assured by large, politically loyal security and military forces; a government apparatus responsive to party direction; and the placement of party members into leadership positions in labor, women's, and other mass organizations.

The "Great Leap Forward" and the Sino-Soviet Split

In 1958, Mao broke with the Soviet model and announced a new economic program, the "Great Leap Forward," aimed at rapidly raising industrial and agricultural production. Giant cooperatives (communes) were formed, and "backyard factories" dotted the Chinese landscape. The results were disastrous. Normal market mechanisms were disrupted, agricultural production fell behind, and China's people exhausted themselves producing what turned out to be shoddy, unsalable goods. Within a year, starvation appeared even in fertile agricultural areas. From 1960 to 1961, the combination of poor planning during the Great Leap Forward and bad weather resulted in one of the deadliest famines in human history.

The already strained Sino-Soviet relationship deteriorated sharply in 1959, when the Soviets started to restrict the flow of scientific and technological information to China. The dispute escalated, and the Soviets withdrew all of their personnel from China in August 1960. In 1960, the Soviets and the Chinese began to have disputes openly in international forums.

The Cultural Revolution

In the early 1960s, State President Liu Shaoqi and his protégé, Party General Secretary Deng Xiaoping, took over direction of the party and adopted pragmatic economic policies at odds with Mao's revolutionary vision. Dissatisfied with China's new direction and his own reduced authority, Party Chairman Mao launched a massive political attack on Liu, Deng, and other pragmatists in the spring of 1966. The new movement, the "Great Proletarian Cultural Revolution," was unprecedented in communist history. For the first time, a section of the Chinese communist leadership sought to rally popular opposition against another leadership group. China was set on a course of political and social anarchy that lasted the better part of a decade.

In the early stages of the Cultural Revolution, Mao and his "closest comrade in arms," National Defense Minister Lin Biao, charged Liu, Deng, and other top party leaders with dragging China back toward capitalism. Radical youth organizations, called Red Guards, attacked party and state organizations at all levels, seeking out leaders who would not bend to the radical wind. In reaction to this turmoil, some local People's Liberation Army (PLA) commanders and other officials maneuvered to outwardly back Mao and the radicals while actually taking steps to rein in local radical activity.

Gradually, Red Guard and other radical activity subsided, and the Chinese political situation stabilized along complex factional lines. The leadership conflict came to a head in September 1971, when Party Vice Chairman and Defense Minister Lin Biao reportedly tried to stage a coup against Mao; Lin Biao allegedly later died in a plane crash in Mongolia.

In the aftermath of the Lin Biao incident, many officials criticized and dismissed during 1966-69 were reinstated. Chief among these was Deng Xiaoping, who reemerged in 1973 and was confirmed in 1975 in the concurrent posts of Politburo Standing Committee member, PLA Chief of Staff, and Vice Premier.

The ideological struggle between more pragmatic, veteran party officials and the radicals re-emerged with a vengeance in late 1975. Mao's wife, Jiang Qing, and three close Cultural Revolution associates (later dubbed the "Gang of Four") launched a media campaign against Deng. In January 1976, Premier Zhou Enlai, a popular political figure, died of cancer. On April 5, Beijing citizens staged a spontaneous demonstration in Tiananmen Square in Zhou's memory, with strong political overtones of support for Deng. The authorities forcibly suppressed the demonstration. Deng was blamed for the disorder and stripped of all official positions, although he retained his party membership.

The Post-Mao Era

Mao's death in September 1976 removed a towering figure from Chinese politics and set off a scramble for succession. Former Minister of Public Security Hua Guofeng was quickly confirmed as Party Chairman and Premier. A month after Mao's death, Hua, backed by the PLA, arrested Jiang Qing and other members of the "Gang of Four." After extensive deliberations, the Chinese Communist Party leadership reinstated Deng Xiaoping to all of his previous posts at the 11th Party Congress in August 1977. Deng then led the effort to place government control in the hands of veteran party officials opposed to the radical excesses of the previous two decades.

The new, pragmatic leadership emphasized economic development and renounced mass political movements. At the pivotal December 1978 Third Plenum (of the 11th Party Congress Central Committee), the leadership adopted economic reform policies aimed at expanding rural income and incentives, encouraging experiments in enterprise autonomy, reducing central planning, and attracting foreign direct investment into China. The plenum also decided to accelerate the pace of legal reform, culminating in the passage of several new legal codes by the National People's Congress in June 1979.

After 1979, the Chinese leadership moved toward more pragmatic positions in almost all fields. The party encouraged artists, writers, and journalists to adopt more critical approaches, although open attacks on party authority were not permitted. In late 1980, Mao's Cultural Revolution was officially proclaimed a catastrophe. Hua Guofeng, a protégé of Mao, was replaced as premier in 1980 by reformist Sichuan party chief Zhao Ziyang and as party General Secretary in 1981 by the even more reformist Communist Youth League chairman Hu Yaobang.

Reform policies brought great improvements in the standard of living, especially for urban workers and for farmers who took advantage of opportunities to diversify crops and establish village industries. Literature and the arts blossomed, and Chinese intellectuals established extensive links with scholars in other countries.

At the same time, however, political dissent as well as social problems such as inflation, urban migration, and prostitution emerged. Although students and intellectuals urged greater reforms, some party elders increasingly questioned the pace and the ultimate goals of the reform program. In December 1986, student demonstrators, taking advantage of the loosening political atmosphere, staged protests against the slow pace of reform, confirming party elders' fear that the current reform program was leading to social instability. Hu Yaobang, a protégé of Deng and a leading advocate of reform, was blamed for the protests and forced to resign as CCP General Secretary in January 1987. Premier Zhao Ziyang was made General Secretary and Li Peng, former Vice Premier and Minister of Electric Power and Water Conservancy, was made Premier.

1989 Student Movement and Tiananmen Square

After Zhao became the party General Secretary, the economic and political reforms he had championed came under increasing attack. His proposal in May 1988 to accelerate price reform led to widespread popular complaints about rampant inflation and gave opponents of rapid reform the opening to call for greater centralization of economic controls and stricter prohibitions against Western influence. This precipitated a political debate, which grew more heated through the winter of 1988-89.

The death of Hu Yaobang on April 15, 1989, coupled with growing economic hardship caused by high inflation, provided the backdrop for a large-scale protest movement by students, intellectuals, and other parts of a disaffected urban population. University students and other citizens camped out in Beijing's Tiananmen Square to mourn Hu's death and to protest against those who would slow reform. Their protests, which grew despite government efforts to contain them, called for an end to official corruption and for defense of freedoms guaranteed by the Chinese constitution. Protests also spread to many other cities, including Shanghai, Chengdu, and Guangzhou.

Martial law was declared on May 20, 1989. Late on June 3 and early on the morning of June 4, military units were brought into Beijing. They used armed force to clear demonstrators from the streets. There are no official estimates of deaths in Beijing, but most observers believe that casualties numbered in the hundreds.

After June 4, while foreign governments expressed horror at the brutal suppression of the demonstrators, the central government eliminated remaining sources of organized opposition, detained large numbers of protesters, and required political reeducation not only for students but also for large numbers of party cadre and government officials.

Following the resurgence of conservatives in the aftermath of June 4, economic reform slowed until given new impetus by Deng Xiaoping's dramatic visit to southern China in early 1992. Deng's renewed push for a market-oriented economy received official sanction at the 14th Party Congress later in the year as a number of younger, reform-minded leaders began their rise to top positions. Deng and his supporters argued that managing the economy in a way that increased living standards should be China's primary policy objective, even if "capitalist" measures were adopted. Subsequent to the visit, the Communist Party Politburo publicly issued an endorsement of Deng's policies of economic openness. Though not completely eschewing political reform, China has consistently placed overwhelming priority on the opening of its economy.

Third Generation of Leaders

Deng's health deteriorated in the years prior to his death in 1997. During that time, President Jiang Zemin and other members of his generation gradually assumed control of the day-to-day functions of government. This "third generation" leadership governed collectively with President Jiang at the center.

In March 1998, Jiang was re-elected President during the 9th National People's Congress. Premier Li Peng was constitutionally required to step down from that post. He was elected to the chairmanship of the National People's Congress. Zhu Rongji was selected to replace Li as Premier.

Fourth Generation of Leaders

In November 2002, the 16th Communist Party Congress elected Hu Jintao, who in 1992 was designated by Deng Xiaoping as the "core" of the fourth generation leaders, the new General Secretary. A new Politburo and Politburo Standing Committee was also elected in November.

In March 2003, General Secretary Hu Jintao was elected President at the 10th National People's Congress. Jiang Zemin retained the chairmanship of the Central Military Commission. At the Fourth Party Plenum in September 2004, Jiang Zemin retired from the Central Military Commission, passing the Chairmanship and control of the People's Liberation Army to President Hu Jintao.

China is firmly committed to economic reform and opening to the outside world. The Chinese leadership has identified reform of state industries and the establishment of a social safety network as government priorities. Government strategies for achieving these goals include large-scale privatization of unprofitable state-owned enterprises and development of a pension system for workers. The leadership has also downsized the government bureaucracy.

The Next 5 Years

The next 5 years represent a critical period in China's existence. To investors and firms, especially following China's accession to the World Trade Organization (WTO) in 2001, China represents a vast market that has yet to be fully tapped and a low-cost base for export-oriented production. Educationally, China is forging ahead as partnerships and exchanges with foreign universities have helped create new research opportunities for its students. The new leadership is also committed to generating greater economic development in the interior and providing more services to those who do not live in China's coastal areas. However, there is still much that needs to change in China. Human rights issues remain a concern among members of the world community, as does continuing proliferation of weapons of mass destruction (WMD)-related materials and technology.

ECONOMY

GDP (2003 est.): $1.4 trillion (exchange rate based).
Per capita GDP (2003 est.): $1,090 (exchange rate based).
GDP real growth rate (2003): 9.1%.
Natural resources: Coal, iron ore, crude oil, mercury, tin, tungsten, antimony, manganese, molybdenum, vanadium, magnetite, aluminum, lead, zinc, uranium, hydropower potential (world's largest).
Agriculture: *Products*--Among the world's largest producers of rice, potatoes, sorghum, peanuts, tea, millet, barley; commercial crops include cotton, other fibers, and oilseeds; produces variety of livestock products.
Industry: *Types*--iron, steel, coal, machinery, light industrial products, armaments, petroleum.
Trade (2003): *Exports*--$438.4 billion: mainly electrical machinery and equipment, power generation equipment, apparel, toys, footwear. *Main partners*--U.S., Hong Kong, Japan, EU, South Korea, Singapore. *Imports*--$412.8 billion: mainly electrical equipment, power generation equipment, petroleum products, chemicals, steel. *Main partners*--Japan, EU, Taiwan, South Korea, U.S., Hong Kong.

Economic Reforms

Since 1979, China has reformed and opened its economy. The Chinese leadership has adopted a more pragmatic perspective on many political and socioeconomic problems, and has reduced the role of ideology in economic policy. China's ongoing economic transformation has had a profound impact not only on China but on the world. The market-oriented reforms China has implemented over the past two decades have unleashed individual initiative and entrepreneurship. The result has been the largest reduction of poverty and one of the fastest increases in income levels ever seen. China today is the sixth-largest economy in the world. It accounted for about 4% of global gross domestic product (GDP) in 2002.

In the 1980s, China tried to combine central planning with market-oriented reforms to increase productivity, living standards, and technological quality without exacerbating inflation, unemployment, and budget deficits. China pursued agricultural reforms, dismantling the commune system and introducing a household-based system that provided peasants greater decision-making in agricultural activities. The government also encouraged nonagricultural activities such as village enterprises in rural areas, and promoted more self-management for state-owned enterprises, increased competition in the marketplace, and facilitated direct contact between Chinese and foreign trading enterprises. China also relied more upon foreign financing and imports.

During the 1980s, these reforms led to average annual rates of growth of 10% in agricultural and industrial output. Rural per capita real income doubled. China became self-sufficient in grain production; rural industries accounted for 23% of agricultural output, helping absorb surplus labor in the countryside. The variety of light industrial and consumer goods increased. Reforms began in the fiscal, financial, banking, price-setting, and labor systems.

By the late 1980s, however, the economy had become overheated with increasing rates of inflation. At the end of 1988, in reaction to a surge of inflation caused by accelerated price reforms, the leadership introduced an austerity program.

China's economy regained momentum in the early 1990s. During a visit to southern China in early 1992, China's paramount leader at the time, Deng Xiaoping, made a series of political pronouncements designed to reinvigorate the process of economic reform. The 14th Party Congress later in the year backed Deng's renewed push for market reforms, stating that China's key task in the 1990s was to create a "socialist market economy." The 10-year development plan for the 1990s stressed continuity in the political system with bolder reform of the economic system.

China's economy grew at an average rate of 10% per year during the period 1990-2001, the highest growth rate in the world. China's gross domestic product (GDP) grew 8% in 2002, and even faster, 9.1%, in 2003, despite the setbacks of the severe acute respiratory syndrome (SARS) outbreak and a sluggish world economy. China's total trade in 2003 surpassed $852 billion, making China the world's fourth-largest trading nation.

Nevertheless, serious imbalances exist behind the spectacular trade performance, high investment flows, and high GDP growth. High numbers of non-performing loans weigh down the state-run banking system. Inefficient state-owned enterprises (SOEs) are still a drag on growth, despite announced plans to sell, merge, or close the vast majority of SOEs.

Social and economic indicators have improved since reforms were launched, but rising inequality is evident between the more highly developed coastal provinces and the less developed, poorer inland regions. According to the Asian Development Bank, about 10.5% of the urban population and 25.5% of the rural population would be classified as poor in 1999.

Following the Chinese Communist Party's Third Plenum, held in October 2003, Chinese legislators unveiled several proposed amendments to the state constitution. One of the most significant was a proposal to provide protection for private property rights. Legislators also indicated there would be a new emphasis on certain aspects of overall government economic policy, including efforts to reduce unemployment (now in the 8-10% range in urban areas), to rebalance income distribution between urban and rural regions, and to maintain economic growth while protecting the environment and improving social equity. The National People's Congress approved the amendments when it met in March 2004.

Agriculture

Roughly half of China's labor force is engaged in agriculture, even though only 10% of the land is suitable for cultivation. China is among the world's largest producers of rice, potatoes, sorghum, millet, barley, peanuts, tea, and pork. Major non-food crops include cotton, other fibers, and oilseeds. Yields are high because of intensive cultivation, but China hopes to further increase agricultural production through improved plant stocks, fertilizers, and technology. Incomes for Chinese farmers are stagnating, leading to an increasing wealth gap between the cities and countryside. Government policies that continue to emphasize grain self-sufficiency and the fact that farmers do not own--and cannot buy or sell--the land they work have contributed to this situation. In addition, inadequate port facilities and lack of warehousing and cold storage facilities impede both domestic and international agricultural trade.

Industry

Major state industries are iron, steel, coal, machine building, light industrial products, armaments, and textiles. These industries have resisted significant management change. The 1999 industrial census revealed that there were 7,930,000 industrial enterprises at the end of 1999 (including small-scale town and village enterprises); total employment in state-owned industrial enterprises was approximately 24 million. High-tech industries are well positioned to take advantage of opportunities created by accession to the WTO. Machinery and consumer products have become China's main exports.

Energy

In 2003, China surpassed Japan to become the second-largest consumer of primary energy, after the United States. China is also the third-largest energy producer in the world, after the United States and Russia. China's electricity consumption is expected to grow by over 4% a year through 2030, which will require more than $2 trillion in electricity infrastructure investment to meet the demand. China expects to add approximately 15,000 megawatts of generating capacity a year, with 20% of that coming from foreign suppliers.

Coal makes up the bulk of China's energy consumption (64% in 2002), and China is the largest producer and consumer of coal in the world. As China's economy continues to grow, China's coal demand is projected to rise significantly. Although coal's share of China's overall energy consumption will fall, coal consumption will continue to rise in absolute terms.

Due in large part to environmental concerns, Beijing would like to shift China's current energy mix toward greater reliance on oil, natural gas, renewable energy, and nuclear power. China has abundant hydroelectric resources; the Three Gorges Dam, for example, will have a total capacity of 18 gigawatts when fully on-line (projected for 2009). In addition, the share of electricity generated by nuclear power is projected to grow from 1% in 2000 to 5% in 2030. But while interest in renewable sources of energy is growing, except for hydropower, their contribution to the overall energy mix is unlikely to rise above 1%-2% in the near future.

Since 1993, China has been a net importer of oil. Net imports are expected to rise to 3.5 million barrels per day by 2010. China is interested in diversifying the sources of its oil imports and has invested in oil fields around the world, particularly in Central Asia. Beijing also plans to increase China's natural gas production, which currently accounts for only 3% of China's total energy consumption. Analysts expect China's consumption of natural gas to more than double by 2010.

Environment

One of the serious negative consequences of China's rapid industrial development has been increased pollution and degradation of natural resources. A 1998 World Health Organization report on air quality in 272 cities worldwide concluded that seven of the world's 10 most polluted cities were in China. According to China's own evaluation, two-thirds of the 338 cities for which air-quality data are available are considered polluted--two-thirds of them moderately or severely so. Respiratory and heart diseases related to air pollution are the leading cause of death in China. Almost all of the nation's rivers are considered polluted to some degree, and half of the population lacks access to clean water. Ninety percent of urban water bodies are severely polluted. Water scarcity also is an issue; for example, severe water scarcity in Northern China is a serious threat to sustained economic growth and has forced the government to begin implementing a large-scale diversion of water from the Yangtze River to northern cities, including Beijing and Tianjin. Acid rain falls on 30% of the country. Various studies estimate pollution costs the Chinese economy 7-10% of GDP each year.

China's leaders are increasingly paying attention to the country's severe environmental problems. In March 1998, the State Environmental Protection Administration (SEPA) was officially upgraded to a ministry-level agency, reflecting the growing importance the Chinese Government places on environmental protection. In recent years, China has strengthened its environmental legislation and made some progress in stemming environmental deterioration. In 1999, China invested more than 1% of GDP in environmental protection, a proportion that will likely increase in coming years. During the 10th Five-Year Plan, China plans to reduce total emissions by 10%. Beijing in particular is investing heavily in pollution control as part of its campaign to host a successful Olympiad in 2008. Some cities have seen improvement in air quality in recent years.

China is an active participant in the climate change talks and other multilateral environmental negotiations, taking environmental challenges seriously but pushing for the developed world to help developing countries to a greater extent. It is a signatory to the Basel Convention governing the transport and disposal of hazardous waste and the Montreal Protocol for the Protection of the Ozone Layer, as well as the Convention on International Trade in Endangered Species and other major environmental agreements.

The question of environmental impacts associated with the Three Gorges Dam project has generated controversy among environmentalists inside and outside China. Critics claim that erosion and silting of the Yangtze River threaten several endangered species, while Chinese officials say the dam will help prevent devastating floods and generate clean hydroelectric power that will enable the region to lower its dependence on coal, thus lessening air pollution.

The United States and China have been engaged in an active program of bilateral environmental cooperation since the mid-1990s, with an emphasis on clean energy technology and the design of effective environmental policy. While both governments view this cooperation positively, China has often compared the U.S. program, which lacks a foreign assistance component, with those of Japan and several European Union (EU) countries that include generous levels of aid.

Science and Technology

Science and technology have always preoccupied China's leaders; indeed, China's political leadership comes almost exclusively from technical backgrounds and has a high regard for science. Deng called it "the first productive force." Distortions in the economy and society created by party rule have severely hurt Chinese science, according to some Chinese science policy experts. The Chinese Academy of

Sciences, modeled on the Soviet system, puts much of China's greatest scientific talent in a large, under-funded apparatus that remains largely isolated from industry, although the reforms of the past decade have begun to address this problem.

Chinese science strategists see China's greatest opportunities in newly emerging fields such as biotechnology and computers, where there is still a chance for China to become a significant player. Most Chinese students who went abroad have not returned, but they have built a dense network of trans-Pacific contacts that will greatly facilitate U.S.-China scientific cooperation in coming years. The United States is often held up as the standard of modernity in China. Indeed, photos of the Space Shuttle often appear in Chinese advertisements as a symbol of advanced technology. China's small but growing space program, which put an astronaut into orbit in October 2003, is a focus of national pride.

The U.S.-China Science and Technology Agreement remains the framework for bilateral cooperation in this field. A 5-year agreement to extend the Science and Technology Agreement was signed in April 2001. There are currently over 30 active protocols under the Agreement, covering cooperation in areas such as marine conservation, renewable energy, and health. Japan and the European Union also have high profile science and technology cooperative relationships with China. Biennial Joint Commission Meetings on Science and Technology bring together policymakers from both sides to coordinate joint science and technology cooperation. Executive Secretaries meetings are held each year to implement specific cooperation programs.

Trade

China's merchandise exports totaled $438.4 billion and imports totaled $412.8 billion in 2003. Its global trade surplus was down 16%, to $25.6 billion. China's primary trading partners include Japan, the EU, the United States, South Korea, Hong Kong, and Taiwan. According to U.S. statistics, China had a trade surplus with the U.S. of $124 billion in 2003.

China has taken important steps to open its foreign trading system and integrate itself into the world trading system. In November 1991, China joined the Asia-Pacific Economic Cooperation (APEC) group, which promotes free trade and cooperation in the economic, trade, investment, and technology spheres. China served as APEC chair in 2001, and Shanghai hosted the annual APEC leaders meeting in October of that year.

China formally joined the WTO in December 2001. As part of this far-reaching trade liberalization agreement, China agreed to lower tariffs and abolish market impediments. Chinese and foreign businessmen, for example, gained the right to import and export on their own, and to sell their products without going through a government middleman. By 2005, average tariff rates on key U.S. agricultural exports will drop from 31% to 14% and on industrial products from 25% to 9%. The agreement also opens up new opportunities for U.S. providers of services like banking, insurance, and telecommunications. China has made significant progress implementing its WTO commitments, but serious concerns remain, particularly in the realm of intellectual property rights protection.

Export growth continues to be a major component supporting China's rapid economic growth. To increase exports, China has pursued policies such as fostering the rapid development of foreign-invested factories, which assemble imported components into consumer goods for export, and liberalizing trading rights.

The United States is one of China's primary suppliers of power generating equipment, aircraft and parts, computers and industrial machinery, raw materials, and chemical and agricultural products. However, U.S. exporters continue to have concerns about fair market access due to strict testing and standards requirements for some imported products. In addition, a lack of transparency in the regulatory process makes it difficult for businesses to plan for changes in the domestic market structure.

Foreign Investment

China's investment climate has changed dramatically in 24 years of reform. In the early 1980s, China restricted foreign investments to export-oriented operations and required foreign investors to form joint-venture partnerships with Chinese firms. Foreign direct investment (FDI) grew quickly during the 1980s, but stalled in late 1989 in the aftermath of Tiananmen. In response, the government introduced legislation and regulations designed to encourage foreigners to invest in high-priority sectors and regions. Since the early 1990s, China has allowed foreign investors to manufacture and sell a wide range of goods on the domestic market, and authorized the establishment of wholly foreign-owned enterprises, now the preferred form of FDI. However, the Chinese government's emphasis on guiding FDI into manufacturing has led to market saturation in some industries, while leaving China's services sectors underdeveloped. China is now one of the leading recipients of FDI in the world, receiving over $53 billion in 2003, for a cumulative total of $501 billion.

As part of China's accession to the World Trade Organization in 2001, China undertook to eliminate certain trade-related investment measures and to open up specified sectors that had previously been closed to foreign investment. New laws, regulations, and administrative measures to implement these commitments are being issued. Major remaining barriers to foreign investment include opaque and inconsistently enforced laws and regulations and the lack of a rules-based legal infrastructure.

Opening to the outside remains central to China's development. Foreign-invested enterprises produce about half of China's exports, and China continues to attract large investment inflows. Foreign exchange reserves totaled over $403 billion in 2003.

U.S.-CHINA RELATIONS

From Liberation to the Shanghai Communiqué

As the PLA armies moved south to complete the communist conquest of China in 1949, the American Embassy followed the Nationalist government headed by Chiang Kai-shek, finally moving to Taipei later that year. U.S. consular officials remained in mainland China. The new P.R.C. Government was hostile to this official American presence, and all U.S. personnel were withdrawn from the mainland in early 1950. Any remaining hope of normalizing relations ended when U.S. and Chinese communist forces fought on opposing sides in the Korean conflict.

Beginning in 1954 and continuing until 1970, the United States and China held 136 meetings at the ambassadorial level, first at Geneva and later at Warsaw. In the late 1960s, U.S. and Chinese political leaders decided that improved bilateral relations were in their common interest. In 1969, the United States initiated measures to relax trade restrictions and other impediments to bilateral contact. On July 15, 1971, President Nixon announced that his Assistant for National Security Affairs, Dr. Henry Kissinger, had made a secret trip to Beijing to initiate direct contact with the Chinese leadership and that he, the President, had been invited to visit China.

In February 1972, President Nixon traveled to Beijing, Hangzhou, and Shanghai. At the conclusion of his trip, the U.S. and Chinese Governments issued the "Shanghai Communiqué," a statement of their foreign policy views. (For the complete text of the Shanghai Communiqué, see the Department of State Bulletin, March 20, 1972)

In the Communiqué, both nations pledged to work toward the full normalization of diplomatic relations. The U.S. acknowledged the Chinese position that all Chinese on both sides of the Taiwan Strait maintain that there is only one China and that Taiwan is part of China. The statement enabled the U.S. and China to temporarily set aside the "crucial question obstructing the normalization of relations"--Taiwan--and to open trade and other contacts.

Liaison Office, 1973-78

In May 1973, in an effort to build toward the establishment of formal diplomatic relations, the U.S. and China established the United States Liaison Office (USLO) in Beijing and a counterpart Chinese office in Washington, DC. In the years between 1973 and 1978, such distinguished Americans as David Bruce, George H.W. Bush, Thomas Gates, and Leonard Woodcock served as chiefs of the USLO with the personal rank of Ambassador.

President Ford visited China in 1975 and reaffirmed the U.S. interest in normalizing relations with Beijing. Shortly after taking office in 1977, President Carter again reaffirmed the interest expressed in the Shanghai Communiqué. The United States and China announced on December 15, 1978, that the two governments would establish diplomatic relations on January 1, 1979.

Normalization

In the Joint Communiqué on the Establishment of Diplomatic Relations dated January 1, 1979, the United States transferred diplomatic recognition from Taipei to Beijing. The U.S. reiterated the Shanghai Communiqué's acknowledgment of the Chinese position that there is only one China and that Taiwan is a part of China; Beijing acknowledged that the American people would continue to carry on commercial, cultural, and other unofficial contacts with the people of Taiwan. The Taiwan Relations Act made the necessary changes in U.S. domestic law to permit such unofficial relations with Taiwan to flourish.

U.S.-China Relations Since Normalization

Vice Premier Deng Xiaoping's January 1979 visit to Washington, DC, initiated a series of important, high-level exchanges, which continued until the spring of 1989. This resulted in many bilateral agreements--especially in the fields of scientific, technological, and cultural interchange and trade relations. Since early 1979, the United States and China have initiated hundreds of joint research projects and cooperative programs under the Agreement on Cooperation in Science and Technology, the largest bilateral program.

On March 1, 1979, the United States and China formally established embassies in Beijing and Washington, D.C. During 1979, outstanding private claims were resolved, and a bilateral trade agreement was concluded. Vice President Walter Mondale reciprocated Vice Premier Deng's visit with an August 1979 trip to China. This visit led to agreements in September 1980 on maritime affairs, civil aviation links, and textile matters, as well as a bilateral consular convention.

As a consequence of high-level and working-level contacts initiated in 1980, U.S. dialogue with China broadened to cover a wide range of issues, including global and regional strategic problems, political-military questions, including arms control, UN and other multilateral organization affairs, and international narcotics matters.

The expanding relationship that followed normalization was threatened in 1981 by Chinese objections to the level of U.S. arms sales to Taiwan. Secretary of State Alexander Haig visited China in June 1981 in an effort to resolve Chinese questions about America's unofficial relations with Taiwan. Eight months of negotiations produced the U.S.-China joint communiqué of August 17, 1982. In this third communiqué, the U.S. stated its intention to reduce gradually the level of arms sales to Taiwan, and the Chinese described as a fundamental policy their effort to strive for a peaceful resolution to the Taiwan question. Meanwhile, Vice President Bush visited China in May 1982.

High-level exchanges continued to be a significant means for developing U.S.-China relations in the 1980s. President Reagan and Premier Zhao Ziyang made reciprocal visits in 1984. In July 1985, President Li Xiannian traveled to the United States, the first such visit by a Chinese head of state. Vice President Bush visited China in October 1985 and opened the U.S. Consulate General in Chengdu, the U.S.'s fourth consular post in China. Further exchanges of cabinet-level officials occurred between 1985-89, capped by President Bush's visit to Beijing in February 1989.

In the period before the June 3-4, 1989 crackdown, a large and growing number of cultural exchange activities undertaken at all levels gave the American and Chinese peoples broad exposure to each other's cultural, artistic, and educational achievements. Numerous Chinese professional and official delegations visited the United States each month. Many of these exchanges continued after Tiananmen.

Bilateral Relations After Tiananmen

Following the Chinese authorities' brutal suppression of demonstrators in June 1989, the U.S. and other governments enacted a number of measures to express their condemnation of China's blatant violation of the basic human rights of its citizens. The U.S. suspended high-level official exchanges with China and

weapons exports from the U.S. to China. The U.S. also imposed a number of economic sanctions. In the summer of 1990, at the G-7 Houston summit, Western nations called for renewed political and economic reforms in China, particularly in the field of human rights.

Tiananmen disrupted the U.S.-China trade relationship, and U.S. investors' interest in China dropped dramatically. The U.S. Government also responded to the political repression by suspending certain trade and investment programs on June 5 and 20, 1989. Some sanctions were legislated; others were executive actions. Examples include:

- The U.S. Trade and Development Agency (TDA)--new activities in China were suspended from June 1989 until January 2001, when then-President Clinton lifted this suspension.
- Overseas Private Insurance Corporation (OPIC)--new activities suspended since June 1989.
- Development Bank Lending/IMF Credits--the United States does not support development bank lending and will not support IMF credits to China except for projects that address basic human needs.

- Munitions List Exports--subject to certain exceptions, no licenses may be issued for the export of any defense article on the U.S. Munitions List. This restriction may be waived upon a presidential national interest determination.

- Arms Imports--import of defense articles from China was banned after the imposition of the ban on arms exports to China. The import ban was subsequently waived by the Administration and reimposed on May 26, 1994. It covers all items on the Bureau of Alcohol, Tobacco, Firearms, and Explosives' Munitions Import List.

In 1996, the P.R.C. conducted military exercises in waters close to Taiwan in an apparent effort at intimidation. The United States dispatched two aircraft carrier battle groups to the region. Subsequently, tensions in the Taiwan Strait diminished, and relations between the U.S. and China have improved, with increased high-level exchanges and progress on numerous bilateral issues, including human rights, nonproliferation, and trade. Former Chinese president Jiang Zemin visited the United States in the fall of 1997, the first state visit to the U.S. by a Chinese president since 1985. In connection with that visit, the two sides reached agreement on implementation of their 1985 agreement on peaceful nuclear cooperation, as well as a number of other issues. Former President Clinton visited China in June 1998. He traveled extensively in China, and direct interaction with the Chinese people included live speeches and a radio show, allowing the President to convey first-hand to the Chinese people a sense of American ideals and values.

Relations between the U.S. and China were severely strained by the tragic accidental bombing of the Chinese Embassy in Belgrade in May 1999. By the end of 1999, relations began to gradually improve. In October 1999, the two sides reached agreement on humanitarian payments for families of those who died and those who were injured as well as payments for damages to respective diplomatic properties in Belgrade and China.

In April 2001, a Chinese F-8 fighter collided with a U.S. EP-3 reconnaissance aircraft flying over international waters south of China. The EP-3 was able to make an emergency landing on China's Hainan Island despite extensive damage; the P.R.C. aircraft crashed with the loss of its pilot. Following extensive negotiations, the crew of the EP-3 was allowed to leave China 11 days later, but the U.S. aircraft was not permitted to depart for another 3 months. Subsequently, the relationship, which had cooled following the incident, gradually improved.

Following the September 11, 2001 terrorist attacks (9-11) in New York City and Washington, DC, China offered strong public support for the war on terrorism and has been an important partner in U.S. counterterrorism efforts. China voted in favor of UN Security Council Resolution 1373, publicly supported the coalition campaign in Afghanistan, and contributed $150 million of bilateral assistance to Afghan reconstruction following the defeat of the Taliban. Shortly after 9-11, the U.S. and China also commenced a counterterrorism dialogue. The third round of that dialogue was held in Beijing in February 2003.

China and the U.S. have also been working closely on regional issues like North Korea. China has stressed its opposition to the D.P.R.K.'s decision to withdraw from the Nuclear Non-Proliferation Treaty, its concerns over North Korea's nuclear capabilities, and its desire for a non-nuclear Korean peninsula. It also voted to refer the D.P.R.K.'s noncompliance with its IAEA obligations to the UN Security Council in New York.

U.S.-China Economic Relations

U.S. direct investment in China covers a wide range of manufacturing sectors, several large hotel projects, restaurant chains, and petrochemicals. U.S. companies have entered agreements establishing more than 20,000 equity joint ventures, contractual joint ventures, and wholly foreign-owned enterprises in China. More than 100 U.S.-based multinationals have projects in China, some with multiple investments. Cumulative U.S. investment in China is valued at $35 billion.

Total two-way trade between China and the U.S. grew from $33 billion in 1992 to over $180 billion in 2003. The United States is China's second-largest trading partner, and China is now the third-largest trading partner for the United States (after Canada and Mexico). U.S. exports to China have been growing more rapidly than to any other market (up 15.3% in 2002 and 28.4% in 2003). U.S. imports from China grew somewhat slower, at 21.7%, but the U.S. trade deficit with China exceeded $124 billion in 2003. Some of the factors that influence the U.S. trade deficit with China include:

- A shift of low-end assembly industries to China from the newly industrialized economies (NIEs) in Asia. China has increasingly become the last link in a long chain of value-added production. Because U.S. trade data attributes the full value of a product to the final assembler, Chinese value-added gets over-counted.
- U.S. demand for labor-intensive goods exceeds domestic output.
- China's restrictive trade practices, which have included an array of barriers to foreign goods and services, often aimed at protecting state-owned enterprises. Under its WTO accession agreement, China is reducing tariffs and eliminating import licensing requirements, as well as addressing other trade barriers.

Source: U.S. Department of State (2004)
http://www.state.gov

Chapter References

Chayet, Neil. 1983. "Law and Morality." Pp. 418-19 in *Life Studies: A Thematic Reader*, edited by D. Cavitch. New York: St. Martin's.

Dunne, John Gregory. 1991. "Law and Disorder in Los Angeles." *The New York Review* (October 10): 23-29.

_____. 1991b. "Law and Disorder in Los Angeles." *The New York Review* (October 24):62-70.

Janofsky, Michael. 1994. "Antismoking Forces at the Barricades? Bring 'em On!" *The New York Times* (April 24):8.

Levine, Dennis B. and William Hoffer. 1991. *Inside Out: An Insider's Account of Wall Street*. New York: Putnam.

Tomashoff, Craig. 1993. "America's Least Wanted Criminals." *Los Angeles Times* (May 10):E1+.

The New Yorker. 1993. "Wrongful Death." (August 16):4-6.

Answers

Concept Application

1. Claims maker	pg. 231
2. White-collar crime	pg. 227
3. Secret deviant	pg. 223
4. Falsely accused	pg. 223
5. Mores	pg. 213

Multiple-Choice

1.	b	pg. 208	14.	c	pg. 223
2.	d	pg. 210	15.	a	pg. 223
3.	c	pg. 213	16.	b	pg. 225
4.	c	pg. 213	17.	c	pg. 227
5.	a	pg. 213	18.	c	pg. 231
6.	b	pg. 214	19.	a	pg. 230
7.	b	pg. 217	20.	a	pg. 237
8.	d	pg. 217	21.	d	pg. 232
9.	d	pg. 219	22.	c	pg. 238
10.	a	pg. 219	23.	c	pg. 244
11.	b	pg. 229	24.	c	pg. 244
12.	c	pg. 220	25.	c	pg. 247
13.	d	pg. 223			

True/False

1.	F	pg. 209
2.	T	pg. 212
3.	T	pg. 219
4.	T	pg. 222
5.	F	pg. 220
6.	F	pg. 223
7.	T	pg. 227
8.	F	pg. 224
9.	T	pg. 239
10.	T	pg. 245
11.	T	pg. 228

Chapter 8

Social Stratification
With Emphasis on the World's Richest and Poorest People

Study Questions

1. As members of the world's richest country, what questions about wealth distribution are we obligated to ask?

2. Why does the social stratification chapter emphasize the world's richest and poorest peoples?

3. What is the connection between social stratification and life chances?

4. What is the connection between place and life chances?

5. What can we say about the distribution of wealth when we compare the income held by the richest 10 percent of a population against the poorest 10 percent?

6. Distinguish between achieved and ascribed characteristics.

7. What does status value mean?

8. What characteristics distinguish a caste from a class system of stratification?

9. Explain the basic dynamics of apartheid.

10. Is the United States a class system? Why or why not?

11. Are caste and class systems distinct types of stratification systems? Explain.

12. In what ways is inequality in the United States systematic?

13. How do functionalists (Davis and Moore) explain stratification? What are some of the shortcomings of their explanations?

14. How do the "functions of poverty" help us to understand whose needs are being met by a system that pays so many so little for their labor?

15. Distinguish between colonialism and neocolonialism.

16. From a world system perspective, how has capitalism come to dominate the global network of economic relationships?

17. Distinguish among core, peripheral, and semiperipheral economies. Give an example of a country that fits each of the three economies.

18. Summarize how Marx approached social class in his writings. Identify three ideas that Marx gave us for approaching social class.

19. How does Max Weber use the concept of social class?

20. How is class ranking complicated by status groups and parties?

21. What general structural changes in the American economy have created an underclass?

22. Who are the members of Responsible Wealth? What principles guide the organization?

Key Concepts

Social stratification	pg. 253
Place	pg. 253
Life chances	pg. 253
Ascribed characteristics	pg. 255
Achieved characteristics	pg. 256
Status value	pg. 256
Caste system	pg. 257
Apartheid	pg. 258
Class system	pg. 257
Social mobility	pg. 259
Vertical mobility	pg. 259
Downward mobility	pg. 259
Upward mobility	pg. 259
Intergenerational mobility	pg. 259
Intragenerational mobility	pg. 259
Class	pg. 273
Negatively privileged	pg. 276
Positively privileged	pg. 276
Urban underclass	pg. 278
Status group	pg. 276
Political parties	pg. 276

Concept Application

Consider the concepts listed below. Match one of more of the concepts with each scenario. Explain your choices.

- ✓ Achieved characteristics
- ✓ Intergenerational mobility
- ✓ Life chances
- ✓ Negatively privileged property class
- ✓ Social stratification
- ✓ Status group
- ✓ Status value
- ✓ Upward mobility
- ✓ Vertical mobility

Scenario 1 "Do blondes have more fun? Social scientists have yet to nail down the answer. But economists now have good reason to believe that blondes make more money—or at least the trim, attractive ones do. New studies show that men and women (with any hair color) who are rated below average in attractiveness by survey interviewers typically earn 10 to 20 percent less than those rated above average.

One is tempted to write off the results as proof that idle econometricians are the Devil's helpers. But the findings from Daniel Hamermesh of the University of Texas and Jeff Biddle of Michigan State are complemented by other research showing that obese women are also at a considerable earnings disadvantage. And they could figure prominently in the very serious business of deciding who is protected by the three-year-old Americans with Disabilities Act" (Passell 1994:C2).

Scenario 2 "The Brinks Hotel was another American symbol in Saigon. It was a bachelor officers' headquarters, an American world that Vietnamese need not enter unless of course it was to clean the rooms or to cook, or to provide some other form of service. It stood high over Saigon and its hovels, a world of Americans eating American food, watching American movies, and just to make sure that there was a sense of home, on the roof terrace there was always a great charcoal grill on which to barbecue thick American steaks flown in especially to that end" (Halberstam 1987:618-19).

Scenario 3 "These children [of people who make enough money to live a privileged life] learn to live with *choices*: more clothes, a wider range of food, a greater number of games and toys, that other boys and girls may never be able to imagine. They learn to grow fond of or resolutely ignore, dolls and more dolls, large dollhouses and all sorts of utensils and furniture to go in them, enough Lego sets to build yet another house for the adults in the family. They learn to take for granted enormous playrooms, filled to the brim with trains, helicopters, boats, punching bags, Monopoly sets.... They learn to assume instruction—not only at school, but at home—for tennis, for swimming, for dancing, for horse riding. And they learn often enough to feel competent at those sports, in control of themselves while playing them, and, not least, able to move smoothly from one to the other" (Coles 1978:26).

Scenario 4 Wanting out is a common ambition in small towns all over America. In 1951, there were three ways to realize it. One was to get a job in the big city—in my case, either Kansas City or St. Louis, at the edges of the imaginable world. At sixteen, I was too young for this, and besides, I had no idea of what I could do.

A second way—chosen by four men from the class ahead of me—was to enlist in a branch of military service or volunteer for the draft. That would get you even farther from home and pile up educational benefits under the GI Bill.

A third alternative to work and military service was just beginning to open up to people—mostly men—of my class and region: college. (Davis 1996:14)

Scenario 5 The deeper message of Edin's book concerns the material hardships that most welfare families still endure. Eight in 10 had severe housing problems. One in six had recently been homeless. One-third had run out of food sometime in the previous year. And conditions didn't really improve for those who appear to have moved up one step to an entry-level job. In examining the budgets of 165 working mothers, Edin found them even more likely than those on welfare to be unable to pay their bills. "I thought they might be the same, but not worse," she says. (DePerle 1997:34)

Applied Research

In *Images of Japanese Society*, Ross E. Mouer and Yoshio Sugimoto (1990) present a multidimensional framework for thinking about stratification. They identify four dimensions of stratification: economic, political, psychological, and information-based, and give examples of rewards associated with each dimension. Select one specific type of reward from the list below and find data from the U.S. Bureau of the Census or other data sources that illustrate patterns of inequality in the United States.

Economic rewards:
- Occupation
- Salary
- Pension
- Benefits
- Environment (quality of surroundings)
- Employment security
- Job safety
- Quality of recreation facilities
- Leisure

Political rewards:
- Influence
- Authority
- Contacts
- Access to guns and tanks
- Control over army or police force
- Votes
- Publicity
- Information and intelligence

Psychological rewards:
- Status
- Prestige
- Honor
- Esteem
- Fame
- Publicity
- Recognition
- Friends
- Conspicuous consumption

Information-based rewards:
- Knowledge
- Specific skills
- Social awareness
- Technical know-how
- Access to information

Practice Test: Multiple-Choice Questions

1. Approximately _______________ people in the world are worth at least $1 million (excluding the value of their homes).
   a. 7.7 million
   b. 20 million
   c. 50.3 million
   d. 100 million

2. The World Bank has identified the 49 poorest countries in the world. Which one of the following is among the five poorest countries in the world?
   a. Mexico
   b. Haiti
   c. Democratic Republic of the Congo
   d. North Korea

3. The systematic process by which individuals, groups, and places are ranked on a scale of social worth is
   a. social stratification.
   b. symbolic stratification.
   c. apartheid.
   d. social structure.

4. A baby born in _______________ has the best chance of surviving its first year of life.
   a. the United States
   b. Sweden
   c. Italy
   d. Japan

5. Consumers living in the highest-income countries account for _____________ percent of total private consumption.
   a. 98
   b. 86
   c. 50
   d. 25

6. If we confine our analysis to the world's 25 wealthiest countries, the inequality gap between the richest 10 percent and poorest 10 percent is greatest in
   a. Japan.
   b. the United States.
   c. Brazil.
   d. South Korea.

7. People assign __________ when they regard some features of a characteristic as more valuable or worthy than other features.
   a. life chances
   b. status value
   c. social class
   d. social stratification

8. In a caste system of social stratification
   a. inequality is not systematic.
   b. there is a systematic connection between ascribed characteristics and life chances.
   c. people can change their class position through hard work.
   d. talent, merit, and ability determine a person's life chances.

9. Apartheid is the Afrikaans word for
   a. discrimination.
   b. separate but equal.
   c. apartness.
   d. racial stratification.

10. Which racial category in South Africa has the best chance of getting to work by car?
    a. white
    b. Indian
    c. mixed or colored
    d. African

11. In comparison to class systems, caste systems of stratification
    a. are extremely rigid.
    b. rank people on the basis of achievements.
    c. have few barriers to social interaction among people from different strata.
    d. allow marriage between people of different strata.

12. A person who changes their class position through marriage, graduation, inheritance, or job promotions is experiencing
    a. vertical mobility.
    b. horizontal mobility.
    c. caste mobility.
    d. downward mobility.

13. Income profiles for households classified as white, black, and Hispanic show that
    a. black and Hispanic households compare favorably with white households.
    b. black and Hispanic households are disproportionately concentrated in lower-income categories.
    c. income distribution is not affected by type of household.
    d. low paying and low prestige occupations are equally distributed in all household types.

14. The fact that income and occupation are connected to race means that we must conclude that the United States is at best
    a. a class system.
    b. a caste system.
    c. a mixture of class and caste system.
    d. nonstratified.

15. In which of the following occupational categories are women most likely to be underrepresented?
    a. child-care
    b. dental assistants
    c. dentists
    d. cleaners and servants

16. In the case of coed baseball, we can also say a caste system is at work because
    a. those who have the most ability and experience (i.e., men) are assigned to the most central positions.
    b. social practices contribute to differences between males' and females' talent and ability.
    c. there is a clear connection between achieved characteristics and position.
    d. biological differences explain differences between males' and females' athletic ability.

17. From a functionalist perspective, social inequality
    a. causes people in the entry-level jobs to work harder.
    b. ensures that the best-qualified people will fill the most demanding positions.
    c. increases the motivation level of all workers.
    d. guarantees that the least-qualified people will not seek the most important jobs.

18. Comparable worth means
    a. that when men and women work in the same firms in the same occupation, they must not be paid differently.
    b. when occupational categories are agreed to be equivalently valuable within a firm, the compensation must be equivalent across those categories.
    c. male and female dominated occupations should be valued equally.
    d. men and women can be paid differently even if they are in the same occupation.

19. The catfood industry is a female-dominated industry while the dogfood industry is male-dominated. Personnel analysts in the catfood industry (who are primarily female) earn $14.00/hour on average. Personnel managers in the dogfood industry (who are primarily male) earn $18.00/hour. This pay difference is related to issues of
    a. pay equity.
    b. comparable worth.
    c. class stratification.
    d. mobility.

20. _______________ is a form of domination in which a foreign power uses its superior military force to impose its political, economic, social, and cultural institutions on an indigenous population with the aim of dominating their resources, labor, and markets.
    a. Neocolonialism
    b. Social stratification
    c. Conflict
    d. Colonialism

21. According to world system theorists, capitalism has come to dominate the world economy because
    a. under this system, governments control economic activities.
    b. it is the only economic system in the world.
    c. of the ways in which capitalists respond to changes in the economy, especially to economic stagnation.
    d. national interests take precedence over corporate interests.

22. The Vietnamese economy is very vulnerable to price fluxuations. For example, the country managed to become the third largest producer of coffee beans when prices were around $3.00 per pound only to see the price drop to less than 50 cents. The vulnerability explains why Vietnam is classified as a _________ economy.
    a. core
    b. peripheral
    c. semiperipheral
    d. middle-income

23. __________________ economies operate on the fringes of the world economy.
    a. Core
    b. Peripheral
    c. Semiperipheral
    d. Service

24. The negatively privileged property classes include all but which one of the following?
    a. completely unskilled persons
    b. those dependent on seasonal employment
    c. those at the bottom of the class system
    d. the bourgeoisie

25. According to Max Weber, persons completely unskilled, lacking property, and dependent on seasonal or sporadic employment constitute the
    a. negatively privileged property class.
    b. ascribed property class.
    c. marketless class.
    d. negatively privileged status group.

26. In the United States, in any given month, an estimated ________ percent of the U.S. population lives in poverty.
    a. 5-8
    b. 13-16
    c. 20-23
    d. 40-43

27. The United Nations calculates that a ____________ levy on the richest 225 people could provide for the base needs of the world's 1.2 billion poorest people.
    a. 4
    b. 10
    c. 15
    d. 25

True/False Questions

T F 1. Since 9-11, global priorities have favored development assistance to the poor.

T F 2. Achieved and ascribed characteristics are clear-cut categories.

T F 3. In a true class system, ascribed characteristics determine one's social class.

T (F) 4. All evidence indicates that the United States possesses a true class system.

(T) F 5. From a functionalist perspective poverty contributes to the stability of society overall.

(T) F 6. Nike sport-shoe is the largest indirect employer in Vietnam.

T (F) 7. Karl Marx wrote *The Communist Manifesto* with Emile Durkheim.

(T) F 8. The world's richest people are concentrated in North America, Western Europe, and China.

Internet Resources Related to Social Stratification

- **Russell Sage Foundation**
  http://www.russellsage.org/
  The Russell Sage Foundation, established in 1907 by Mrs. Olivia Sage, is dedicated to the "improvement of social and living conditions in the United States." Among other programs, the Russell Sage Foundation supports research related to the decline in demand among advanced economies for low-skill workers.

- **Northwestern University Policy Research**
  http://www.northwestern.edu/ipr/publications/newsletter/index.html
  The Institute for Policy Research (formerly the Center for Urban Affairs and Policy Research) publishes a newsletter which addresses a variety of social issues. A few titles from the articles are "IPR Policy Briefing Explores Mass Incarceration in the U.S." and "Do Women Have What it Takes to Lead? Study shows women are just as effective leaders as men."

Internet Resources Related to Poor and to the World's Richest and Poorest Countries

- **The Bible on the Poor**
  http://www.zompist.com/meetthepoor.html
  "The Bible contains more than 300 verses on the poor, social justice, and God's deep concern for both. This page contains a wide sample of them, and some reflections. It's aimed at anyone who takes the Bible seriously."

- **Debt Relief Under the Heavily Indebted Poor Countries (HIPC) Initiative**
  http://www.imf.org/external/np/exr/facts/hipc.htm
  The HIPC Initiative was intended to resolve the debt problems of the most heavily-indebted poor countries (originally 41 countries, mostly in Africa) with total debt nearing $200 billion. The 600 million people living in these countries survive an average of 7 years less than citizens in other developing countries, with half living on less than $1 per day. Money freed up by debt relief must be used for sustainable development, so that the countries will not again face unmanageable debts and their people can exit from extreme poverty.

- **Paying the Price**
  http://www.oxfam.org.uk/what_we_do/issues/debt_aid/mdgs_price.htm
  "A new report from international agency Oxfam today reveals that 45 million more children will die needlessly by 2015, because rich countries are failing to provide the necessary resources to overcome poverty."

- **Rich World, Poor World**
  http://www.cgdev.org/Research/?TopicID=39
  "The Center for Global Development is dedicated to reducing global poverty and inequality through policy-oriented research and active engagement on development issues with the policy community and the public. A principal focus of the Center's work is the policies of the United States and other industrial countries that affect development prospects in poor countries." This site gives more information on CBD research and publications on issues such as *Education and the Developing World, Global Trade, Jobs and Labor Standards* and *Global HIV/AIDS and the Developing World.*

- **Ranking the Rich**
  http://www.foreignpolicy.com/story/files/story2540.php
  The second annual CGD/FP Commitment to Development Index ranks 21 rich nations on how their aid, trade, investment, migration, environment, security, and technology policies help poor countries.

Background Notes: Luxembourg: World Richest Country as Measured by Per Capita GDP

PEOPLE

Nationality: *Noun*--Luxembourger(s). *Adjective*--Luxembourg.
Population (July 2004 est.): 462,690.
Annual growth rate (2004 est.): 1.28%.
Ethnic groups: Celtic base with French and German blend; also guest workers from Portugal, Italy, France, and other European countries.
Religion: Historically Roman Catholic. Luxembourg law forbids the collection of data on religious practices.
Official languages: Luxembourgish, French, and German; English is widely spoken.
Education: *Years compulsory*--9. *Attendance*--100%. *Literacy*--100%.
Health: *Life expectancy*--avg. 78 years; males 75 years, females 82 years. *Infant mortality rate*--4.88/1000.

Work force (2004, 293,670): *Services*--27%; *agriculture*--1%; *industry*--13%; *government and social services*--22%; *financial services* --28%; *construction*--8%.
Unemployment (September 2004 est.): 4.2%.

HISTORY

The national language of Luxembourg is Luxembourgish, a blend of Dutch, old German, and Frankish elements. The official language of the civil service, law, and parliament is French, although criminal and legal debates are conducted partly in Luxembourgish and police case files are recorded in German. German is the primary language of the press. French and German are taught in the schools, with German spoken mainly at the primary level and French at the secondary level.

After 400 years of domination by various European nations, Luxembourg was granted the status of Grand Duchy by the Congress of Vienna on June 9, 1815. Although Luxembourg considers 1835 (Treaty of London) to be its year of independence, it was not granted political autonomy until 1839 under King William I of the Netherlands, who also was the Grand Duke of Luxembourg. In 1867, Luxembourg was recognized as independent and guaranteed perpetual neutrality. After being occupied by Germany in both World Wars, Luxembourg abandoned neutrality and became a charter member of the North Atlantic Treaty Organization (NATO) in 1949.

The present sovereign, Grand Duke Henri, succeeded his father, Grand Duke Jean, on October 7, 2000. Grand Duke Jean announced his decision to abdicate in December 1999, after a 35-year reign.

ECONOMY

GDP (2003 est.): $25.01 billion (22.39 billion EUR)
Annual growth rate (2003): 2.9%
Per capita income (2003 est.): $55,100.
Inflation rate (September 2004): 1.9%.
Natural resources: Iron ore.
Agriculture (0.5% of GNP): *Products*--dairy products, corn, wine. *Arable land*--49%.
Services (2003): 83.1%.
Industry (16.3% of GNP): *Types*--chemicals, steel.
Trade (2003 est.): *Exports*-- $10.138 billion: steel, plastics, rubber and processed wood products. Major markets--Germany, Belgium, France, and Asia. *Imports*--$13.506 billion: minerals, including iron ore, coal, and petroleum products; mechanical and electrical equipment, transportation equipment, scrap metal. *Major suppliers*--other EU countries (esp. Belgium, France, and Germany).

Although Luxembourg is aptly described as the "Green Heart of Europe" in tourist literature, its pastoral land coexists with a highly industrialized and export-intensive economy. Luxembourg enjoys a degree of economic prosperity almost unique among industrialized democracies.

In 1876, English metallurgist Sidney Thomas invented a refining process that led to the development of the steel industry in Luxembourg and the founding of the Arbed company in 1911. In 2001, Arbed merged with Aceralia and Usinor to form Arcelor, the world's largest steel producer, which is headquartered in Luxembourg. The iron and steel industry, located along the French border, is the most important single sector of the economy. In 2002 steel accounted for 27% of all exports (excluding services), 30% of industrial employment, and 3.8% of the work force.

There has been a relative decline in the steel sector, offset by Luxembourg's emergence as a financial center. The financial sector in 2002 made up more than 35% of Luxembourg's gross domestic product. Banking is especially important. In 2004 (October), there were 167 banks in Luxembourg, with 23,300 employees. Political stability, good communications, easy access to other European centers, skilled multilingual staff, and a tradition of banking secrecy have contributed to the growth of the financial sector. Germany accounts for the largest single grouping of banks, with Scandinavian, Japanese, and major U.S. banks also heavily represented. Total banking assets exceeded $676 billion at the end of August 2004. Approximately 14,000 holding companies are established in Luxembourg.

Government policies promote the development of Luxembourg as an audiovisual and communications center. Radio-Television-Luxembourg is Europe's premier private radio and television broadcaster. The government-backed Luxembourg satellite company Société Européenne des Satellites (SES) was created in 1986 to install and operate a satellite telecommunications system for transmission of television programs throughout Europe. The first SES "ASTRA" satellite, a 16-channel RCA 4000, was launched by Ariane rocket in December 1988. SES presently operates 13 satellites. ASTRA 1H is the most advanced satellite with a return channel capacity in the Ka band frequency range enabling two-way satellite communications directly to users' terminals.

Luxembourg offers a favorable climate to foreign investment. Successive governments have effectively attracted new investment in medium, light, and high-tech industry. Incentives cover taxes, construction, and plant equipment. The recent European Union (EU) directive on services supplied electronically has caused a number of companies to look to Luxembourg, with its relatively low value-added tax (VAT) rates, as a possible location for directing their European operations. U.S. firms are among the most prominent foreign investors, producing tires (Goodyear), chemicals (Dupont), glass (Guardian Industries), and a wide range of industrial equipment. The Department of Commerce's Bureau of Economic Analysis reports that total U.S. direct investment in Luxembourg (on a historical cost basis) was nearly $67 billion at the end of 2003. Foreign direct investment (FDI) data for Luxembourg must be interpreted cautiously, however, because of Luxembourg's role in financial intermediation, particularly involving Luxembourg-based holding companies.

Labor relations have been peaceful since the 1930s. Most industrial workers are organized by unions linked to one of the major political parties. Representatives of business, unions, and government participate in the conduct of major labor negotiations.

Foreign investors often cite Luxembourg's labor relations as a primary reason for locating in the Grand Duchy. Unemployment in 2004 has averaged approximately 4.2%.

Luxembourg's small but productive agricultural sector provides employment for less than 2% of the work force. Most farmers are engaged in dairy and meat production. Vineyards in the Moselle Valley annually produce about 15 million liters of dry white wine, most of which is consumed locally.

Luxembourg's trade account has run a persistent deficit over the last decade, but the country enjoys an overall balance-of-payment surplus, due to revenues from financial services. Government finances are strong, and budgets are normally in surplus.

U.S.-LUXEMBOURG RELATIONS

The United States and Luxembourg have traditionally enjoyed a strong relationship, expressed both bilaterally and through common membership in NATO, the Organization for Economic Cooperation and Development (OECD) and the Organization for Security and Cooperation in Europe (OSCE). More than 5,000 American soldiers, including Gen. George S. Patton, are buried at the American Military Cemetery near the capital.

Background Notes: East Timor: World's Poorest Country as Measured by Per Capita GDP

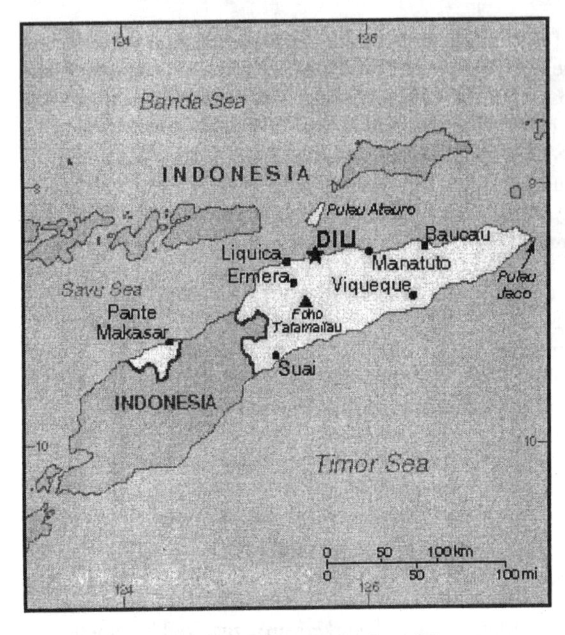

PEOPLE

Nationality: *Noun*--Timorese; *adjective*--Timorese.
Population (2004 est.): 850,000.
Ethnic groups: Maubere.
Religion: Catholic 98% Languages: Portuguese, Tetum (official languages); English, Bahasa Indonesia (working languages).
Education: *Literacy*--41%.
Health: *Life expectancy*--49.5 years. *Mortality rate* (under 5)--126. per 1,000 live births.

East Timor is located in southeastern Asia, on the southernmost edge of the Indonesian archipelago, northwest of Australia. The country includes the eastern half of Timor island as well as the Oecussi enclave in the northwest portion of Indonesian West Timor, and the islands of Atauro and Jako. The mixed Malay and Pacific Islander culture of the Timorese people reflects the geography of the country on the border of those two cultural areas. Portuguese influence during the centuries of colonial rule resulted in a substantial majority of the population identifying itself as Roman Catholic. Some of those who consider themselves Catholic practice a mixed form of religion that includes local animist customs. As a result of the colonial education system and the 23-year Indonesian occupation, approximately 17% of Timorese speak Portuguese and 63% speak Bahasa Indonesia. Tetum, the most common of the local languages, is spoken by approximately 91% of the population. Mambae, Kemak, and Fataluku are also widely spoken. This linguistic diversity is enshrined in the country's constitution which designates Portuguese and Tetum as official languages and English and Bahasa Indonesia as working languages.

HISTORY

Portuguese and Dutch traders made the first western contact with East Timor in the early 16th century. Sandalwood and spice traders, as well as missionaries, maintained sporadic contact with the island until 1642, when the Portuguese moved into Timor in strength. The Portuguese and the Dutch, based at the western end of the island in Kupang, battled for influence until the present-day borders were agreed to by the colonial powers in 1906. Imperial Japan occupied East Timor from 1942-45. Portugal resumed colonial authority over East Timor in 1945 after the Japanese defeat in World War II.

Following a military coup in Lisbon in April 1974, Portugal began a rapid and disorganized decolonization process in most of its overseas territories, including East Timor. Political tensions—

exacerbated by Indonesian involvement--heated up, and on August 11, 1975, the Timorese Democratic Union Party (UDT) launched a coup d'état in Dili. The putsch was followed by a brief but bloody civil war in which the Revolutionary Front for an Independent East Timor (FRETILIN) pushed UDT forces into Indonesian West Timor. Shortly after the FRETILIN victory in late September, Indonesian forces began incursions into East Timor. On October 16, five journalists from Australia, Britain, and New Zealand were murdered in the East Timorese town of Balibo shortly after they had filmed regular Indonesian army troops invading East Timorese territory. On November 28, FRETILIN declared East Timor an independent state, and Indonesia responded by launching a full-scale military invasion on December 7. On December 22, 1975 the UN Security Council called on Indonesia to withdraw its troops from East Timor.

Declaring a provisional government made up of Timorese allies on January 13, 1976, the Indonesian Government said it was acting to forestall civil strife in East Timor and to prevent the consolidation of power by the FRETILIN party. The Indonesians claimed that FRETILIN was communist in nature, while the party's leadership described itself as social democratic. Coming on the heels of the communist victories in Vietnam, Cambodia, and Laos, the Indonesian claims were accepted by many in the West. Major powers also had little incentive to confront Indonesia over a territory seen as peripheral to their security interests. Nonetheless, the widespread popular support shown for the guerilla resistance launched by the Timorese made clear that the Indonesian occupation was not welcome. The Timorese were not permitted to determine their own political fate via a free vote, and the Indonesian occupation was never recognized by the United Nations.

Indonesian occupation of Timor was initially characterized by a program of brutal military repression. Beginning in the late 1980s, however,the occupation was increasingly characterized by programs to win the "hearts-and-minds" of the Timorese through the use of economic development assistance and job creation while maintaining a strict policy of political repression, although serious human rights violations – such as the 1991 Santa Cruz massacre -- continued. Estimates of the number of Timorese who lost their lives to violence and hunger during the Indonesian occupation range from 100,000 to 250,000. On January 27, 1999, Indonesian President B.J. Habibie announced his government's desire to hold a referendum in which the people of East Timor would chose between autonomy within Indonesia and independence. Under an agreement among the United Nations, Portugal, and Indonesia, the referendum was held on August 30, 1999. When the results were announced on September 4--78% voted for independence with a 98.6% turnout--Timorese militias organized and supported by the Indonesian military commenced a large-scale, scorched-earth campaign of retribution. While pro-independence FALINTIL guerillas remained cantoned in UN-supervised camps, the militia killed approximately 1,300 Timorese and forcibly pushed 300,000 people into West Timor as refugees. The majority of the country's infrastructure, including homes, irrigation systems, water supply systems, and schools, and nearly 100% of the country's electrical grid were destroyed. On September 20, 1999 the Australian-led peacekeeping troops of the International Force for East Timor (INTERFET) deployed to the country, bringing the violence to an end.

ECONOMY

GDP (2002 est.): $400 million.
GDP per capita (nominal): $497.
GDP composition by sector: Services 57%, agriculture 25%, industry 17%.
Industry: *Types*--coffee, oil and natural gas.
Trade: *Exports*--coffee, oil and natural gas. *Major markets*--Australia, Europe, Japan, United States. *Imports*--basic manufactures, commodities. *Major sources*--Australia, Europe, Indonesia, Japan, United States.

As the poorest nation in Asia, East Timor must overcome formidable challenges. Basic income, health, and literacy indicators are among the lowest in Asia. Severe shortages of trained and competent personnel to staff newly established executive, legislative, and judicial institutions hinder progress. Rural areas, lacking in infrastructure and resources, remain brutally poor, and the relatively few urban areas cannot provide adequate jobs for the country's growing labor force. Many cities, including the country's second largest, Baucau, do not have routine electrical service. Rural families' access to electricity and clean water is very limited. While anticipated revenues from offshore oil and gas reserves, expected to begin in late 2004, offer great hope for the country, effective use of those resources will require a major transformation of the country's current human and institutional infrastructure. Meanwhile, as those substantial revenues come on line, foreign assistance levels--now standing at among the highest worldwide on a per capita basis--will likely taper off.

East Timor has made significant progress in a number of areas since independence. It has become a full-fledged member of the international community, joining the United Nations, the International Monetary Fund, the World Bank, and the Asian Development Bank (ADB). It is surviving the massive exodus of UN personnel, equipment and resources, and has effected a relatively smooth transition to Timorese control of the government and its administration. It produced a National Development Plan, and its Constituent Assembly has transitioned into a National Parliament that has commenced reviewing and passing legislation. A nascent legal system has been put into place and efforts are underway to put in place the institutions required to protect human rights, rebuild the economy, create employment opportunities, and reestablish essential public services.

U.S.-EAST TIMOR RELATIONS

East Timor maintains an embassy in Washington DC, as well as a Permanent Mission in New York at the United Nations. The United States has a large bilateral development assistance program, $22.5 million in 2004, and also contributes funds as a major member of a number of multilateral agencies such as the Asian Development Bank and World Bank. The U.S. Peace Corps has an active program in East Timor.

Source: US Department of State (2004)
http://www.state.gov

Chapter References

Coles, Robert. 1978. *Privileged Ones: The Well-Off and the Rich in America*. Boston: Little, Brown.

Davis, Robert Murray. 1996. *A Lower-Middle Class Education*. Norman, OK: University of Oklahoma Press.

Halberstam, David. 1987. *The Best and the Brightest*. New York Penguin.

Mouer, Ross E. and Yoshio Sugimoto. 1990. *Images of Japanese Society: A Study in the Social Construction of Reality*. New York: Routledge, Chapman & Hall.

Passell, Peter. 1994. "Economic Scene." *The New York Times* (January 27):C2.

Answers

Concept Application

1. Status value; Life chances; Ascribed characteristics — pg. 252; 276; 256
2. Status group — pg. 276
3. Life chances; Social stratification — pg. 253
4. Upward mobility, intergenerational mobility, vertical mobility — pg. 259
5. Negatively privileged property class — pg. 276

Multiple-Choice

1.	a	pg. 251
2.	c	pg. 253
3.	a	pg. 253
4.	b	pg. 254
5.	b	pg. 255
6.	b	pg. 255
7.	b	pg. 256
8.	b	pg. 258
9.	c	pg. 258
10.	a	pg. 259
11.	a	pg. 258
12.	a	pg. 259
13.	b	pg. 261
14.	c	pg. 262
15.	c	pg. 262
16.	b	pg. 263
17.	b	pg. 264
18.	b	pg. 266
19.	b	pg. 266
20.	d	pg. 270
21.	c	pg. 270
22.	b	pg. 272
23.	b	pg. 272
24.	d	pg. 276
25.	a	pg. 276
26.	b	pg. 277
27.	a	pg. 278

True/False

1.	F	pg. 252
2.	F	pg. 256
3.	F	pg. 258
4.	F	pg. 262
5.	T	pg. 267
6.	T	pg. 272
7.	F	pg. 273
8.	F	pg. 279

Chapter 9

Race and Ethnic Classification
With Emphasis on the Peopling of the United States (A Global Story)

Study Questions

1. Explain how the peopling of the United States is a global story.

2. Why is it important to focus on the U.S. system of racial and ethnic classification?

3. Name the 6 official racial categories as designated by the U.S. Census Bureau. How did the system of racial classification change in 1997?

4. What are the shortcoming associated with assigning people to clear-cut racial categories?

5. How does the U.S. Census Bureau define Hispanic? Do those classified as Hispanic see themselves as "Hispanic"? Why or why not?

6. Define chance, choice, and context. How is race a product of these factors?

7. How is ethnicity a product of chance, choice, and context?

8. Why are people of Middle Eastern and Arab ancestry classified as "white"?

9. Why is the 2003 Census population brief on Arab populations within the United States historic?

10. What percentage of the U.S. population is foreign-born? Who are the foreign-born?

11. How has race and ethnicity been connected to U.S. immigration policy?

12. Immigration has always inspired debate in the U.S. Why?

13. What are minority groups? What are the essential characteristics of all minority groups?

14. Distinguish between absorption assimilation and melting pot assimilation.

15. According to Gordon, which level of absorption assimilation is most important or most critical?

16. What are racist ideologies? Give at least two examples showing how racist ideologies are used to justify one group's domination over another.

17. What are the reasons black athletes dominate some sports such as basketball and are virtually invisible in others?

18. What is a stereotype? How are the stereotypes perpetuated and reinforced?

19. According to Robert K. Merton, what is the relationship between prejudice and discrimination?

20. Distinguish between individual discrimination and institutional discrimination. Give examples.

21. What is race-thinking? List six strategies for keeping race-thinking in check (These six strategies are also ways of eliminating the negative consequences of racial classification).

22. What is a stigma? How is this concept relevant to issues of race and ethnicity?

23. What are mixed contacts? How do stigmas dominate the course of interaction between the stigmatized and normals?

24. How do the stigmatized respond to people who treat them as members of a category?

Key Concepts

Race as a product of	
Chance	pg. 292
Context	pg. 292
Choice	pg. 292
Ethnicity	pg. 293
Involuntary ethnicity	pg. 294
Selective forgetting	pg. 293
Foreign-born	pg. 296
Native (of the United States)	pg. 296

Minority group	pg. 301
Assimilation	pg. 302
Absorption assimilation	pg. 302
Melting pot assimilation	pg. 302
Segregation	pg. 303
Involuntary minorities	pg. 305
Voluntary minorities	pg. 307
Ideology (racist)	pg. 305
Prejudice	pg. 307
Stereotypes	pg. 307
Selective perception	pg. 308
Discrimination	pg. 311
Nonprejudiced nondiscriminators	pg. 311
Prejudiced nondiscriminators	pg. 313
Nonprejudiced discriminators	pg. 312
Prejudiced discriminators	pg. 316
Individual discrimination	pg. 317
Institutionalized discrimination	pg. 317
Stigma	pg. 318
Mixed contacts	pg. 320
Race thinking	pg. 326

Concept Application

Consider the concepts listed below. Match one or more of the concepts with each scenario. Explain your choices.

- ✓ Institutionalized discrimination
- ✓ Involuntary minorities
- ✓ Melting pot assimilation
- ✓ Mixed contacts
- ✓ Selective perception
- ✓ Stereotypes
- ✓ Stigma

Scenario 1 "In the book, an American Asian woman finds her White beau attractive because he is from Connecticut, not Canton. He is tall and lanky; he does not have skinny arms like her brothers and father. He is commanding and gets what he wants. Asian men, however, are not depicted as commanding but as arrogant and chauvinistic. My Asian father has never treated my mother arrogantly. He is not short or uncommunicative, either. My father is tall with broad shoulders, a physical attribute inherited by both my brother and me. We have his strong jawline, too. And I have a dimple on my chin like actor Kirk Douglas. An American Asian woman acquaintance made a comment that my brother and I were unlike 'typical' Asian men because we are tall and muscular. Her own brother is tall and muscular! It gets worse. Two strangers from Latin America, on two separate occasions, asked me if I was 'mixed.' Both refused to believe that I was 100% Asian because I did not fit their stereotype of what an Asian should look like. One even referred to my 'big' eyes" (Wang 1994:20).

Scenario 2 "…[I]ntegration can be seen as a two-way process in which the dominant and subordinate sectors interact to forge a new entity, in much the same way as different paints in a bucket. Under integration, the best elements of both the majority and minority culture are merged into a single and coherent national framework across a range of practices, including intermarriage and education" (Fleras and Elliott 1992:62).

Scenario 3 Finally, the category "Native American" is an artifice of the colonial collision. It is composed of multiple socio-cultural groups who share a colonial history as Indians. They, too, were marginalized in and excluded from full and equal participation in mainstream American institutions and practices. First, there was the military conquest of the Native Americans and their subsequent removal to reservations. But, almost from their first interactions, Native Americans sought education from the United States government. In more than one-quarter of the approximately four-hundred treaties entered into by the United States government between 1778 and 1871, education was one of the specific services Native Americans requested in exchange for their lands. But in the formalized education provided by the United States, Native American students were forced to embrace Western ideas and culture, whose price was the repression and denial of their own cultures. Many students were forced into a cultural no-man's land where they remained torn between two worlds. Most students simply dropped out of the system (Hogue 1996:9).

Scenario 4 The Tuskegee study began in 1932, when the Public Health Service (and later the Centers for Disease Control) decided to follow 400 black men with syphilis without treating them. The subjects, who were recruited from churches and clinics throughout the South, were told only that they had "bad blood" (Stryker 1997: E4).

Scenario 5 As soon as I walked into the students' center, I knew I'd gone to the wrong place. Just about everyone there looked really ethnic-African American, Asian, Native American, Latino. And there I was, this white-looking guy. A few other students looked kind of white, too, but at least their names tags made up for it: last names like "Chan" or "Lee" or "Wong." What's my last name? Jewish. Great.

I stood around feeling really out of place until this other student began talking to me. He was African American. "So what are you?" he asked me, right away. I was relieved to tell him my mom was Chinese, like I was explaining myself. "Oh, OK, yeah, you can sort of see it," he said, after eyeing me carefully. "But would you look at some of the guys here? I don't know what they're supposed to be." I left a little later and never went back (Hess 1997).

Applied Research

Choose one of the followings books: *The Color of Water: A Black Man's Tribute to His White Mother* by James McBride, *Member of the Club* by Lawrence Otis Graham, *Showing Our Colors: Afro-German Women Speak Out* edited by May Opitz, Katharina Oguntoye and Dagmar Schultz, *Black Indians: A Hidden Heritage* by William Loren Katz, *Border Lands* by Gloria Anqaldua or *Life on the Color Line* by Gregory Howard Williams. Write a book review describing how the book you have chosen supports the idea that race classification as practiced in the U.S. makes no sense.

Practice Test: Multiple-Choice Questions

1. The peopling of the United States is one of the great dramas of human history. It involved all but which one of the following?
   a. the conquest of Native peoples.
   b. the annexation of Mexican territory.
   c. an influx of involuntary immigrants.
   d. a long-standing tradition of open-immigration policies.

2. In the U.S., there are ____________ race categories if we include single and multiple-race responses.
   a. 63
   b. 103
   c. 213
   d. 13

3. Under the U.S. system of racial classification people with ancestors from Pakistan and Siberia are expected to identify as
   a. Asian.
   b. Native American.
   c. Black.
   d. White.

4. The label Hispanic is confusing because it forces people to identify themselves with conquistadors and settlers from ____________.
   a. Britain.
   b. Brazil.
   c. Spain.
   d. Portugal.

5. ____________ is the larger social setting in which racial and ethnic categories are recognized, constricted, and challenged.
   a. Chance
   b. Context
   c. Choice
   d. Conscious

6. According to the information presented in Chapter 9 (Race and Ethnicity) we can conclude that race is
   a. an inherited trait (like hair color).
   b. a biological fact.
   c. a result of the system of racial classification.
   d. a scientific fact.

7. At one time in the United States people could go into a "white" church if they could run a comb through their hair without it snagging. This situation speaks to the importance of __________ in determining race.
   a. chance
   b. choice
   c. context
   d. understanding

8. Africans forced to immigrate as enslaved people from hundreds of cultures emerged from their experiences as a single _____________ category.
   a. racial
   b. ancestry
   c. ethnic
   d. language

9. The U.S. Bureau of the Census has explored the possibility of creating a new racial category for people of _______________ ancestry.
   a. German
   b. Arab-Middle Eastern
   c. Spanish
   d. Brazilian

10. The racial classification rule in the United States requiring people to choose one racial identity was officially dismantled in
    a. 1900.
    b. 1930.
    c. 1950.
    d. 1997.

11. ___________________ is a situation in which a dominant group defines some subgroup of people in racial or ethnic terms, thereby forcing that subgroup to become, appear, or feel more ethnic than they might otherwise be.
    a. Designated ethnicity
    b. Ethnicity
    c. Foreign ethnicity
    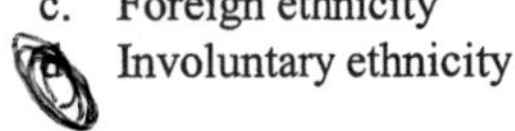
    d. Involuntary ethnicity

12. Which region of the world is home to the largest percentage of foreign-born living in the United States?
    a. Latin America
    b. Asia
    c. Europe
    d. Africa

13. Under the Immigration Act of 1924, a quota system set numerical limits on immigration based on national origin. Immigrants from which region of the world were <u>most</u> affected by this Act?
    a. western Europe
    b. southern, central, and eastern Europe

    c. Asia
    d. Latin America

14. There have been many changes to U.S. immigration law since 9-11. Which one of the following is such a change?
    a. The Department of Defense is home to the Bureau of Citizenship and Immigration Services.
    b. Nationals from all foreign countries must register with the Department of Homeland Security.
    c. International visitors with minor visa problems can no longer enter the United States at the discretion of border and airport inspectors.
    d. Foreign students cannot attend medical school in the United States.

15. "When I use checks, credit cards, or cash, I can count on my skin color not to work against the appearance that I am financially reliable." This statement illustrates an example of
    a. privileges that members of dominant groups enjoy and take for granted.
    b. privileges that members of dominant groups have *earned.*
    c. adaptation to dominant culture.
    d. the workings of capitalist societies.

16. Which one of the following countries has a minority group within its borders that is a numerical majority?
    a. South Africa
    b. Canada
    c. the United States
    d. Israel

17. In __________ assimilation, members of a minority ethnic or racial group adapt to the ways of the dominant group.
    a. melting pot
    b. absorption
    c. involuntary
    d. voluntary

18. Melting pot assimilation
    a. is a one-sided process in which a minority group is absorbed into the dominant culture.
    b. exists when a minority group identifies with the dominant culture.
    c. occurs when a minority group procreates with those of a dominant group.
    d. is a new process of cultural blending in which the groups involved accept new behaviors and values from one another.

19. The concept of melting pot assimilation can be applied to the experiences of
    a. the various African ethnic groups brought to the United States as slaves.
    b. the various Hispanic groups that have settled in the United States.
    c. international labor migrants in Germany.
    d. East and West Germans after the fall of the Berlin Wall.

20. One possible explanation for black-white differences in tipping can be traced to the fact that
    a. blacks deliberately leave poor tips.
    b. blacks are angry at whites for slavery.
    c. until recently, blacks have not been allowed to work as waiters and waitresses.
    d. blacks are exercising reverse discrimination.

21. Which one of the following theories best explains why black male athletes dominate the sport of basketball?
    a. The black male athletes' ability can be traced to the fact that slaveowners bred his ancestors to be strong.
    b. The black male athlete has an extra muscle in his leg.
    c. Blacks are just better athletes, plain and simple.
    d. Black athletes' energy is channeled toward money-making sports such as basketball.

22. Bill notes that "black" athletes dominate the sport of basketball and uses that as evidence of natural leaping ability. At the same time he does not use the same kind of logic to explain why "white" athletes dominate gymnastics. Bill is guilty of
    a. selective perception.
    b. assimilation.
    c. institutionalized discrimination.
    d. non-prejudiced discrimination.

23. The distinction between prejudice and discrimination is that
    a. prejudice is a behavior; discrimination is an attitude.
    b. prejudice is an attitude; discrimination is a behavior.
    c. prejudice is an ideology; discrimination is an attitude.
    d. prejudice is an attitude; discrimination is an ideology.

24. Jerome admits that "whenever someone would break out the inevitable 'black joke,' I would be angered but I didn't want to jeopardize my own social standing by speaking up." Jerome's behavior and conflicting attitude represents that of a/an
    a. nonprejudiced nondiscriminator.
    b. nonprejudiced discriminator.
    c. prejudiced nondiscriminator.
    d. prejudiced discriminator.

25. Laws and practices designed with the clear intention of keeping members of outgroups in subordinate positions fall under the category of
    a. individual discrimination.
    b. institutionalized discrimination.
    c. stigma.
    d. mixed contacts.

26. According to sociologist Erving Goffman, the very anticipation of contact can cause normals and stigmatized to try to avoid one another. This is because the two parties
    a. wish to resist the social pressures pushing them to interact with one another.
    b. wish to avoid discomfort, rejection, and suspicions they encounter from people in the other group.
    c. believe they cannot form a relationship that matches the "ideal" kind of relationship portrayed in the media.
    d. have experienced negative reactions from everyone they have encountered in the other group.

True/False Questions

T F 1. About twenty percent of the U.S. population claims membership in two or more races.

T F 2. Physical boundaries separating one racial category from another are clear and definite.

T F 3. Individual choice regarding race is constrained by chance and context.

T F 4. People simplify complex ethnic backgrounds through selective forgetting.

T F 5. Since 1920, five of the top 10 countries from which people have immigrated to the United States are in Western Europe.

T F 6. Someone born on American soil to two parents with illegal status is considered a U.S. citizen.

T F 7. In apartheid South Africa, the whites were minorities (as defined by Wirth).

T F 8. The melting post concept of assimilation helps explain how the various African ethnic groups imported to the U.S. as slaves became one racial group.

T F 9. The Equal Opportunity Survey commissioned by the U.S. Armed Forces found that whites were the only racial/ethnic group who experienced no discrimination.

T F 10. The Jim Crow Laws mandated busing as a means of ending segregation.

Internet Resources Related to Race

- **Interracial Voice**
  http://www.webcom.com/~intvoice
  Interracial Voice is an electronic publication issued every other month. It "advocates universal recognition of mixed-race individuals as constituting a separate 'racial' entity and supports the initiative to establish a multiracial category on the 2000 census."

- **The Population Reference Bureau**
  http://www.prb.org
  The PRB "provides data, data interpretation, and assistance with key demographic terms to print and broadcast media worldwide." It website includes a section that focuses on race and ethnicity, with special emphasis on reports that focus on race/ethnicity and life chances. Examples of reports include "U.S. Fertility Rates Higher Among Minorities," "Racial and Ethnic Differences in U.S. Mortality," "Homeownership Rates Divide Racial and Ethnic Groups," and "1 Million Arab Americans in the United States."

Internet Resources Related to Racial and Ethnic Categories

- **U.S. Bureau of the Census**
  http://www.census.gov/population/www/socdemo/race.html
  Every ten years the U.S. Bureau of the Census counts and categorizes people living in the United States by race. This website offers immediate access to census brief and other reports on the six official racial categories along with the 57 two or more race categories.

Statistical Profile: Race and Ethnicity in the United States

The Two or More Races Population: 2000

Census 2000 showed that the United States population on April 1, 2000, was 281.4 million. Of the total, 6.8 million people, or 2.4 percent, reported[1] more than one race. Census 2000 asked separate questions on race and Hispanic or Latino origin. Hispanics who reported more than one race are included in the two or more races population.

The term "two or more races" refers to people who chose more than one of the six race categories. These individuals are referred to as the *two or more races* population, or as the population that reported *more than one race.*

Data on race has been collected since the first U.S. decennial census in 1790. Census 2000 was the first decennial census that allowed individuals to self-identify with more than one race.

The question on race was changed in Census 2000

For Census 2000, the question on race was asked of every individual living in the United States and responses reflect self-identification. Respondents were asked to report the race or races they considered themselves and other members of their households to be.

[1] In this report, the term "reported" is used to refer to the answers provided by respondents, as well as responses assigned during the editing and imputation processes.

The question on race for Census 2000 was different from the one for the 1990 census in several ways. Most significantly, respondents were given the option of selecting one or more race categories to indicate their racial identities.

The Census 2000 question on race included 15 separate response categories and 3 areas where respondents could write in a more specific race. The response categories and write-in answers were combined to create the five standard Office of Management and Budget race categories plus the Census Bureau category of "Some other race." The six race categories include:

- White
- Black or African American
- American Indian and Alaska Native
- Asian
- Native Hawaiian and Other Pacific Islander
- Some other race

Table 1: Total Population by Number of Races Reported: 2000

(For information on confidentiality protection, nonsampling error, and definitions, see www.census.gov/prod/cen2000/doc/pl94-171.pdf)

Number of races	Number	Percent of total population	Percent of total Two or more races population
Total Population	281,421,906	100.00	(x)
One race	274,595,678	97.6	(x)
Two or more races	6,826,228	2.4	100.0
Two races	6,368,075	2.3	93.3
Three races	410,285	0.1	6.0
Four races	38,408	-	0.6
Five races	8,637	-	0.1
Six races	823	-	-

- Percentage rounds to 0.0.
X Not applicable.
Source: U.S. Census Bureau, Census 2000 Redistricting Data (Public Law 94-171) Summary File, Table PL1.

The data collected by Census 2000 on race can be divided into two broad categories: the race *alone* population and the *two or more races* population.

People who responded to the question on race by indicating only one race are referred to as the race *alone* population, or the group who reported *only one* race. For example, respondents who marked only the White category on the census questionnaire would be included in the White *alone* population.

Individuals who chose more than one of the six race categories are referred to as the *two or more races* population, or as the group who reported *more than one race*. For example, respondents who reported they were "White ***and*** Black or African American" or "White ***and*** American Indian and Alaska Native ***and*** Asian" would be included in the *two or more races* category.

Census 2000 provides a snapshot of the two or more races population.

Table 1 shows the number and percentage by number of races reported. In the total population, 6.8 million people, or 2.4 percent, reported more than one race. Of the total two or more races population, the overwhelming majority (93 percent) reported exactly two races. An additional 6 percent reported three races, and 1 percent reported four or more races.

THE GEOGRAPHIC DISTRIBUTION OF THE TWO OR MORE RACES POPULATION

The majority of the two or more races population lived in the West.

According to Census 2000, of the total two or more races population,
40 percent lived in the West,
27 percent lived in the South,
18 percent lived in the Northeast, and
15 percent lived in the Midwest.

Nearly two-thirds of all people who reported more than one race lived in just ten states.

The ten states with the largest two or more races populations in 2000 were California, New York, Texas, Florida, Hawaii, Illinois, New Jersey, Washington, Michigan, and Ohio. Combined, these states represented 64 percent of the total two or more races population. These states contained 49 percent of the total population. Three states had two or more races populations greater than one-half million: California was the only state with a two or more races population greater than one million, followed by New York with 590,000, and Texas with 515,000. These three states accounted for 40 percent of the total two or more races population.

There were fourteen states where the two or more races population exceeded the U.S. rate of 2.4 percent, led by the western states of Hawaii (21 percent), followed at a distance by Alaska (5.4 percent), California (4.7 percent), and the southern state of Oklahoma (4.5 percent). The other ten states included the western states of Arizona, Colorado, Nevada, New Mexico, Oregon, and Washington; the northeastern states of New Jersey, New York, and Rhode Island; and the southern state of Texas. No midwestern state had greater than 2.4 percent of its population reporting more than one race. Four states — California, Hawaii, New York, and Washington — were represented in the top ten states for both number and percent reporting more than one race. There were five states where the two or more races population represented 1.0 percent or less of the total population: Alabama, Maine, Mississippi, South Carolina, and West Virginia.

Table 2: Ten Largest Places in Total Population and in Two or More Races Population: 2000

(For information on confidentiality protection, nonsampling error, and definitions, see www.census.gov/prod/cen2000/doc/pl94-171.pdf)

Place	Total population		Two or more races population		
	Rank	Number	Rank	Number	Percen t
New York, NY	1	8,008,278	1	393,959	4.9
Los Angeles, CA	2	3,694, 820	2	191.288	5.2
Chicago, IL	3	2,896,016	3	84,437	2.9
Houston, TX	4	1,953,631	4	61,478	3.1
Philadelphia, PA	5	1,517,50	10	33,574	2.2
Phoenix, AZ	6	1,321,045	8	43,276	3.3
San Diego, CA	7	1,223,400	5	59,081	4.8
Dallas, TX	8	1,188,580	12	32,351	2.7
San Antonio, TX	9	1,144,646	9	41.871	3.7
Detroit, MI	10	951,270	18	22,041	2.3
Honolulu, HI	46	371,657	6	55,474	14.9
San Jose, CA	11	894,943	7	45,062	5.0

Source: U.S. Census Bureau, Census 2000 Redistricting Data (Public Law 94-171) Summary File, Table PL1.

The places with the largest two or more races populations were New York and Los Angeles.

Census 2000 showed that, of all places in the United States with populations of 100,000 or more, New York with nearly 400,000, and Los Angeles with nearly 200,000, had the largest two or more races populations (see Table 2). These places were also the two largest places in the United States. Four other places (Chicago, Houston, San Diego, and Honolulu) had two or more races populations greater than 50,000.

Although Dallas and Detroit were among the ten largest places in the United States, they did not place among the ten largest places with two or more races populations. Honolulu and San Jose had the sixth and seventh largest two or more races populations.

None of the ten largest places in total population ranked among the ten largest places by percent reporting more than one race. Los Angeles (ranked 30th by percent), New York (ranked 42nd), and San Diego (ranked 45th) came closest, with about 5 percent of all respondents reporting more than one race.

Among places of 100,000 or more population, the highest proportion of more than one race reporting was in Honolulu, with 15 percent. But places with populations between 100,000 and 200,000 tended to have higher proportions of two or more races populations than places with greater than 200,000 population. All ten places with the highest proportion of two or more races had over 6 percent of their population reporting more than one race. Eight of the places were in the West, and two were in the Northeast.

ADDITIONAL FINDINGS ON THE TWO OR MORE RACES POPULATION

What proportion of respondents reporting more than one race also reported a Hispanic origin?

The Office of Management and Budget defines Hispanic or Latino as "a person of Cuban, Mexican, Puerto Rican, South or Central American, or other Spanish culture or origin, regardless of race." In data collection and presentation, federal agencies use two ethnicities: "Hispanic or Latino" and "Not Hispanic or Latino." Race and ethnicity are considered two separate and distinct concepts by the federal system. Hispanics may be of any race, and people in the two or more races population can be Hispanic or not Hispanic.

According to Census 2000, about one in three people in the two or more races population also reported as Hispanic. About 6 percent of Hispanics reported more than one race, in contrast to 2 percent of non-Hispanics.

Which race was most likely to be in combination with one or more other races?
The total two or more races population in Census 2000 was 6.8 million, or about 2.4 percent of the total population. But the percent reporting more than one race varied by race. The White population and the Black or African American population had the lowest percentages reporting more than one race. Of the 216.9 million respondents who reported White alone or in combination, 2.5 percent, or 5.5 million, reported White as well as at least one other race. Similarly, of the 36.4 million individuals who reported Black or African American alone or in combination, 4.8 percent, or 1.8 million, reported Black or African American as well as at least one other race.

The Asian population, and the some other race population had somewhat higher percentages reporting more than one race. Of the 11.9 million individuals who reported Asian alone or in combination, 13.9 percent, or 1.6 million, reported Asian as well as at least one other race. Similarly, of the 18.5 million individuals who reported some other race alone or in combination, 17.1 percent, or 3.2 million, reported some other race as well as at least one other race.

The American Indian and Alaska Native population, and the Native Hawaiian and Other Pacific Islander population had the highest percentages reporting more than one race. Of the 4.1 million individuals who reported American Indian and Alaska Native alone or in combination, 39.9 percent, or 1.6 million, reported American Indian and Alaska Native as well as at least one other race. Similarly, of the 874,000 individuals who reported Native Hawaiian and Other Pacific Islander alone or in combination, 54.4 percent, or 476,000, reported Native Hawaiian and Other Pacific Islander as well as at least one other race.

How was the two or more races population distributed across each race group?
In Census 2000, respondents were able to report more than one race, meaning that a combination of two, three, four, five, or six races could be reported. The races reported by individuals who identified with more than one race varied. About four-fifths of these responses included "White" as one of the reported races. About half included "Some other race." "Black or African American," "American Indian and Alaska Native," and "Asian" were reported in about one-fourth of all responses. "Native Hawaiian and Other Pacific Islander" was reported in less than one-tenth of all responses.

Sources: The Two or More Races Population: 2000
http://www.census.gov/population/www/socdemo/race.html

Chapter References

Fleras, Augi and Jean Leonard Elliott. 1992. *Multiculturalism in Canada*. Scarborough, Ontario: Nelson Canada.

Hess, Demian. 1997. "But You Don't Look Chinese" Interracial Voice. http://www.webcom.com/~intvoice/hess1.html.

Hogue, W. Lawrence. 1996. *Race Modernity, Postmodernity*. Albany, NY: State University of New York.

Stryker, Jeff. 1997. "Tuskegee's Long Arm Still Touches a Nerve." *The New York Times* (April 13):4E.

Wang, George. 1994. "A Few Good Images." *Interrace* (June/July):20-21.

Answers

Concept Application

1. Selective perception; Stigma; Mixed contacts; Stereotypes	pg. 308; 318; 320; 307
2. Melting pot assimilation	pg. 303
3. Involuntary minorities	pg. 303
4. Institutionalized discrimination	pg. 317
5. Mixed contacts	pg. 320

Multiple-Choice

1. d	pg. 286	14. c	pg. 300
2. a	pg. 289	15. a	pg. 302
3. a	pg. 290	16. a	pg. 301
4. c	pg. 291	17. b	pg. 302
5. b	pg. 292	18. d	pg. 303
6. c	pg. 292	19. a	pg. 305
7. c	pg. 292	20. c	pg. 306
8. a	pg. 294	21. d	pg. 310
9. b	pg. 295	22. a	pg. 308
10. d	pg. 288	23. b	pg. 311
11. d	pg. 294	24. b	pg. 312
12. a	pg. 299	25. b	pg. 318
13. c	pg. 299	26. b	pg. 321

True/False

1. F	pg. 289
2. F	pg. 288
3. T	pg. 292
4. T	pg. 293
5. T	pg. 300
6. T	pg. 296
7. F	pg. 301
8. T	pg. 305
9. F	pg. 312
10. F	pg. 306

Chapter 10
Gender
With Emphasis on American Samoa

Study Questions

1. Why is American Samoa the focus of a chapter on gender?

2. Define gender. Why do sociologists find the concept of "gender" useful?

3. Is male/female a clear-cut biological distinction? Why or why not?

4. What is gender polarization? Give an example.

5. How are gender-schematic decisions and gender polarization related? Give an example of how gender-schematic decisions affect educational choices and sexual desire.

6. How did Mead describe feeling rules among female adolescents in Samoa and the United States?

7. How did Derek Freeman view Mead's research on Samoan adolescents? In what ways do Mead's findings fit her "purpose"?

8. Explain the following statement: "People of the same sex vary in the extent to which they meet their society's gender expectations."

9. What are *fà-afafines*? What characteristics of Samoan society support this gender blurring?

10. Give some examples showing how socialization operates to teach people society's gender expectations.

11. What socialization mechanisms are at work in American Samoa to encourage interest and success in football among males?

12. What does the commercialization of gender ideals mean? Give at least one example.

13. How do situational theorists explain social and economic differences between men and women? Give at least two examples.

14. How does one's position in the social structure channel behavior in stereotypically male or female directions?

15. How would structural theorists explain the different conclusions Mead and Freeman drew about Samoan society?

16. What is sexist ideology? How is it reflected in social institutions?

17. How is sexist ideology reflected in military policy toward homosexuals?

18. Name five areas of women's lives over which the state may choose to exercise control.

19. How does an awareness of gender issues and the goal of sexual equality strengthen the institution of the family?

Key Concepts

Sex	pg. 332
Primary sex characteristics	pg. 332
Intersexed	pg. 332
Transsexuals	pg. 333
Secondary sex characteristics	pg. 334
Gender	pg. 334
Masculinity	pg. 334
Femininity	pg. 334
Gender polarization	pg. 336
Gender schematic	pg. 338
Commercialization of gender ideals	pg. 350
Social emotion	pg. 338
Aspirational doll	pg. 347
Structural constraints	pg. 351
Evolutionary view [of sex differences]	pg. 355
Ethgender	pg. 358

Concept Application

Consider the concepts listed below. Match one or more of the concepts with each scenario. Explain your choices.

- ✓ Femininity (or feminine characteristics)
- ✓ Gender
- ✓ Gender polarization
- ✓ Structural constraints
- ✓ Intersexuals

Scenario 1 "Women were widely excluded from the jury service until a few decades ago, many years after it was no longer permissible to exclude blacks as a group, and it was not until 1975 that the Supreme Court ruled that states had to maintain a representative jury pool that included women" (Greenhouse 1994:A10).

Scenario 2 High-heeled shoes are still meant predominantly for posing in, as Miss America does in her swimsuit. She keeps her legs together, one knee gently bent. Pictures of women in bathing suits with heeled legs astride make a more up-to-date, but not necessarily a more feminist, statement.

High heels have never been made for comfort or for ease of movement. Their first wearers spoke of themselves as "mounted" or "propped" upon them; they were strictly court wear, and constituted proof that one intended no physical exertion, and need make none.

The Chinese had long known footwear that had the same effect, with wooden pillars under the arch of each shoe, so that wearers required one or even two servants to help them totter along. Women had their feet deformed, by binding, into tiny, almost useless fists, which were shod in embroidered bootees: men got out of the thought of these an unconscionable thrill. (Visser 1994:38)

Scenario 3 "Women in professional jobs have workplace issues like the glass ceiling and the mommy track. But now there is one for secretaries: rug-ranking. 'If the secretary's pay is based on her boss's status, not on the content of her job, that's rug-ranking—treating her as a perk like the size of his office or the quality of the carpet on his floor,' said N. Elizabeth Fried, a labor consultant based in Dublin, Ohio. 'Secretaries are the only ones in the corporate world who pay is directly linked to the boss. Instead of a career path of their own, most secretaries have had a hitch-your-wagon-to-a-star reward system" (Lewin 1994:A1).

Scenario 4 "Family work was structured around gender and age…Women were responsible for the farmyard economy of milking, rearing of young animals, poultry, butter making, and frequently the cultivation of vegetables as well…Animal husbandry, the buying and selling of animals, most fieldwork (e.g. plowing, burrowing) and structural yardwork (e.g., building, repairing) requiring heavy effort, was undertaken by the male 'farmer'." (O'Hara 2001)

Scenario 5 "[There] are rare cases in which babies are born whose sexual gender is ambiguous or indeterminate…Sexually ambiguous infants, who either appear to be female but are biologically male or appear to be male but are biologically female are sometimes called pseudohermaphrodites." (Scarboro 1991:339)

Applied Research

Use the NewsTracker service or the internet (http://nt.excite.com) to monitor articles on gender issues. Monitor the articles for one to two weeks.

Practice Test: Multiple-Choice Questions

1. From a sociological point of view, *Coming of Age in American Samoa* is significant because Mead used Samoa as a vehicle to explore and critique ____________________ in America.
   a. female adolescence
   b. primary sex characteristics
   c. secondary sex characteristics
   d. homophobia

2. ________ determines a baby's sex.
   a. Predestination
   b. Chance
   c. The male biological parent
   d. The female biological parent

3. Spanish hurdler Maria Jose Martinez Patino lost her right to compete in amateur and Olympic events because she failed the sex tests. Martinez challenged that decision. In the end the IAAF ruled that Martinez
   a. was a "man."
   b. was intersexed.
   c. would have to compete with men in the future.
   d. possessed no special advantages over other female competitors.

4. __________ are considered secondary sex characteristics.
   a. Reproductive organs
   b. Distribution patterns of facial and body hair
   c. Chromosomes
   d. Hormones

5. When the painter Paul Gauguin visited Tahiti in 1891 he emphasized that there is "something virile in the women and something feminine in the men." His observations suggest that
   a. Tahiti was a backward society.
   b. homosexuality was acceptable in that society.
   c. there is a fixed line separating maleness from femaleness.
   d. the United States (and Europe) has made women into artificial creations.

6. Before the Christianization of Samoa, the transition from boyhood to manhood was accompanied by
   a. a religious ceremony that lasted for 3 days.
   b. a time of isolation in the wilderness.
   c. separation from females for one year.
   d. a long, painful process of body tattooing from the waist to below the knees.

7. An elementary school student is asked: "How would your life be different if you were a member of the opposite sex?" he replies: "I would have to shave my whole body." His response implies that he thinks in terms of
   a. feminist principles.
   b. gender polarization.
   c. a biological model of sex differences.
   d. gender convergence.

8. A man decides, conscious or unconsciously, that the woman he dates must be shorter and younger than he is. This decision can be classified as
   a. gender-schematic.
   b. natural.
   c. biologically-based.
   d. feminist.

9. Which college major listed below is the most female-dominated?
   a. library science.
   b. parks and recreation.
   c. theological studies.
   d. psychology.

10. Which one of the following statements reflects Mead's assessment of sexual activity among Samoans?
    a. Sexual activity is reserved for important relationships.
    b. Homosexuality is strictly forbidden.
    c. Virginity is highly valued.
    d. Adolescent sexual feelings and energies are not just reserved for opposite-sex relationships.

11. One might argue that Mead "needed" to find a society that contrasted sharply with the United States. In particular, she needed to find a society
    a. where females were valued for something other than physical appearance.
    b. where men and women were equally valued.
    c. where women were knowledgeable about the human body and could express sexual desire free of constraints.
    d. where women had the same opportunities as men.

12. The closest word we have to *fà-afafines* in American society is
    a. males.
    b. transvestite.
    c. transsexual.
    d. intersexed.

13. The two largest employers in American Samoa are
    a. the military and banks.
    b. tourism and tuna canneries.
    c. the government and tuna canneries.
    d. GM and McDonald's.

14. Which statement best reflects the "socialization perspective" on gender differences?
    a. A person's position in the social structure can channel his or her behavior in a stereotypical male or female direction.
    b. There is a close correspondence between primary sex characteristics and athletic ability.
    c. An undetermined but significant portion of male-female differences are products of the ways in which males and females are treated.
    d. Differences between men and women can be traced to their daily work experiences.

15. About one in every ____________ American Samoan male high school graduates leave the island to play football in the U.S.
    a. 50
    b. 20
    c. 8
    d. 4

16. Mattel markets "Barbie" as an/a
    a. aspirational doll.
    b. doll with traditional values.
    c. doll with feminist values.
    d. intersexed toy.

17. The theme of __________ runs through sociologist Renee R. Anspach's research on physicians and nurses working in neonatal intensive care units.
    a. socialization
    b. situational constraints
    c. sexist ideology
    d. reverse discrimination

18. Because Margaret Mead was a women she was denied access to village chiefs and council meetings. Derek Freeman was made an honorary village chief, putting him in a favorable position to do research. These different experiences are examples of
    a. structural constraints.
    b. ideologies.
    c. selective perceptions.
    d. ethngenders.

19. The evolutionary view of sex differences is one example of
    a. a situational constraint.
    b. sexist ideology.
    c. socialization theory.
    d. a sociological perspective.

20. A U.S. Department of Defense directive states that homosexuality is incompatible with military service. This directive is grounded in
    a. socialization theory.
    b. an understanding of situational constraints.
    c. the scientific method.
    d. sexist ideology.

21. Approximately __________ percent of U.S. Senators are women.
    a. 5
    b. 13
    c. 25
    d. 50

22. The focus of the suit filed by the abandoned Filipinos is
    a. establish whether the sexual relationships between servicemen and local women was consensual.
    b. the support of an estimated 8,600 children fathered by U.S. servicemen.
    c. to establish the Navy's direct role in the local bar and sex industry.
    d. to keep the military base in the Philippines.

23. In her studies of male-female body language, Mills found
    a. male body language includes many affiliative clues.
    b. female body language is characterized by serious facial expressions.
    c. females create an overall impression of power, dominance, and high status.
    d. women tend to constrict their arms and legs, sit in attentive upright postures, and lower their eyes frequently.

24. _______________ percent of veterans in the United States are males.
    a. 50
    b. 75
    c. 84
    d. 96

25. The Grameen Bank has 1,175 branches and an estimated 2.4 million borrowers living in 41,000 villages. About 60 percent of borrowers live in
    a. the United States.
    b. India.
    c. Bangladesh
    d. Latin America.

True/False Questions

T 1. *Coming of Age in Samoa* became one of the twentieth century's most influential books.

F 2. Gender is a biologically-based distinction.

T 3. The biological father's contribution of an X or a Y chromosome determines the baby's sex.

T 4. Margaret Mead portrayed Samoan adolescence as a painless, stress free, and untroubled time.

F 5. As a group, males have a longer life expectancy than females.

T 6. The *fà-afafines* in American Samoa imitate popular foreign female vocalists, such as Britney Spears or Madonna.

T 7. Transvestitism was practiced in pre-Christian Samoa.

8. In the United States there is a commercial product on the market to improve almost every female body part or body function.

9. Compared to men, women are channeled disproportionately into lower-paying, dead-end jobs.

10. The fertility rate of women on public assistance is higher than the fertility rate of women not on such assistance.

11. The U.S. military has discharged more than 9,000 servicemen and women under its "Don't Ask, Don't Tell Policy."

Internet Resources Related to Gender

- **Department of Labor: Women's Bureau**
  http://www.dol.gov/wb/welcome.html
  If you are interested in women's labor statistics and issues the Women's Bureau is an important resource. Particularly useful information is available under "Media Releases" and "Fact Sheets About Women [and Men] in the Workplace".

- **International Foundation for Gender Education**
  http://www.ifge.org/index.php
  This site is full of information on all gender types such as transsexuals, transgender, crossdresser and more. Information includes latest news, forums, book reviews, films, and the *Transgender Tapestry Magazine*.

Internet Resources Related to American Samoa

- **Samoa News**
  http://www.samoanews.com/
  This website posts the most significant stories published in the daily edition of *Samoa News* as well as headline news from past issues.

American Samoa: Background Notes

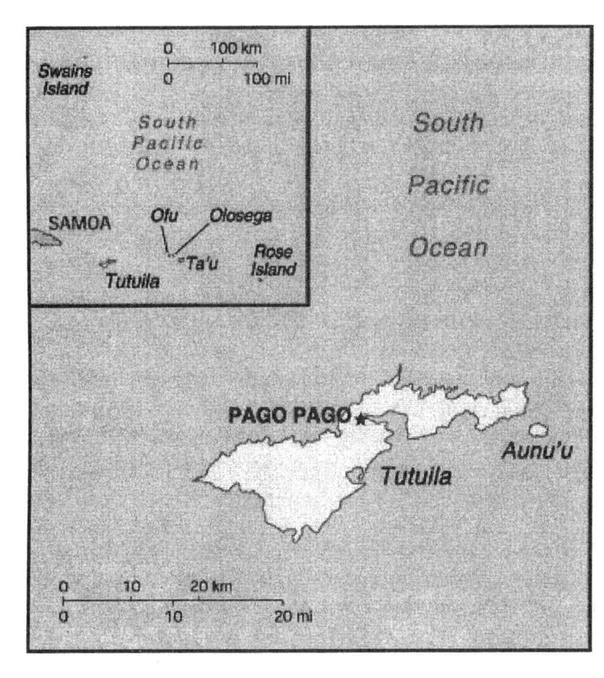

Background

Settled as early as 1000 B. C., Samoa was "discovered" by European explorers in the 18th century. International rivalries in the latter half of the 19th century were settled by an 1899 treaty in which Germany and the US divided the Samoan archipelago. The US formally occupied its portion - a smaller group of eastern islands with the excellent harbor of Pago Pago - the following year.

Area

Total area: 199 sq km
Land area: 199 sq km
Comparative area: Slightly larger than Washington, D.C.
Natural resources: Pumice, pumicite

Environment

Current issues: Limited natural fresh water resources; the water division of the government has spent substantial funds in the past few years to improve water catchments and pipelines.

People

Population: 57,902 (July 2004 est.)

Age structure: *0-14 years:* 36.6% (male 10,983; female 10,208)
15-64 years: 60.3% (male 18,010; female 16,933)
65 years and over: 3.1% (male 699; female 1,069) (2004 est.)

Population growth rate: 0.04% (2004 est.)

Birth rate: 24.46 births/1,000 population (2004 est.)

Death rate: 3.39 deaths/1,000 population (2004 est.)

Net migration rate: -20.71 migrant(s)/1,000 population (2004 est.)

Sex ratio: *at birth:* 1.06 male(s)/female
under 15 years: 1.08 male(s)/female
15-64 years: 1.06 male(s)/female
65 years and over: 0.65 male(s)/female
total population: 1.05 male(s)/female (2004 est.)

Infant mortality rate: 9.48 deaths/1,000 live births

Life expectancy at birth: *total population:* 75.62 years
male: 72.05 years
female: 79.41 years (2004 est.)

Total fertility rate: 3.41 children born/woman (2004 est.)

Economy

Economic overview: This is a traditional Polynesian economy in which more than 90% of the land is communally owned. Economic activity is strongly linked to the US, with which American Samoa conducts most of its foreign trade. Tuna fishing and tuna processing plants are the backbone of the private sector, with canned tuna the primary export. Transfers from the US Government add substantially to American Samoa's economic well-being. Attempts by the government to develop a larger and broader economy are restrained by Samoa's remote location, its limited transportation, and its devastating hurricanes. Tourism is a promising developing sector.

GDP - per capita: purchasing power parity - $8,000 (2000 est.)

GDP - composition by sector: *agriculture:* NA%
industry: NA%
services: NA%

Sources: The World Factbook page on American Samoa (2004)
http://www.cia.gov/cia/publications/factbook/

Chapter References

Greenhouse, Linda. 1994. "High Court Bars Sex As Standard in Picking Jurors." *The New York Times* (April 20):A1+.

Lemann, Nicholas. 1996. "High in the Lower Depth." *The New York Review* (December 19):19-21.

Lewin, Tamar. 1994. "As the Boss Goes, So Goes the Secretary: Is It Bias?" *The New York Times* (March 17):A1+.

O'Hara, Patricia. 2001. "Divisions of Labour on Irish Family Farms." Pp. 270-279 in *Gender in Cross-Cultural Perspective* edited by C.B. Brettell and C.F. Sarent. Uper Saddle River, NJ: Prentice-Hall.

Scarboro, Allen. 1991. "Sexual Ambiguity" Pp. 339-340 in *Women's Studies Encyclopedia* edited by H. Tierney. New York: Peter Bedrick Books.

Visser, Margaret. 1994. *The Way We Are*. Boston: Faber and Faber.

Answers

Concept Application

1. Structural constraints pg. 351
2. Femininity pg. 334
3. Structural constraints pg. 351
4. Gender polarization pg. 336
5. Intersexuals pg. 332

Multiple-Choice

1. a pg. 330
2. c pg. 332
3. d pg. 332
4. b pg. 334
5. d pg. 334
6. d pg. 335
7. b pg. 336
8. a pg. 338
9. a pg. 338
10. d pg. 339
11. c pg. 341
12. b pg. 344
13. c pg. 346
14. c pg. 346
15. c pg. 348
16. a pg. 347
17. b pg. 352
18. a pg. 354
19. b pg. 355
20. d pg. 355
21. b pg. 358
22. b pg. 364
23. d pg. 356
24. d pg. 361
25. c pg. 362

True/False

1. T pg. 331
2. F pg. 334
3. T pg. 332
4. T pg. 341
5. F pg. 338
6. T pg. 344
7. F pg. 344
8. T pg. 350
9. T pg. 351
10. F pg. 360
11. T pg. 359

Chapter 11

Economics and Politics
With Emphasis on Iraq

Study Questions

1. Why focus on Iraq in conjunction with the topics of economics and politics?

2. Define economic system. Name three revolutions that have shaped economic systems.

3. Why is the domestication of plants and animals and the invention of the scratch plow considered revolutionary?

4. Name one of the most fundamental features of the Industrial Revolutions. Why is this feature fundamental?

5. What do we mean when we say the Industrial Revolution cannot be separated from European colonization? Use oil as an example.

6. What is the Information Revolution?

7. What characteristics distinguish a capitalist economic system from a socialist one?

8. From a world system perspective, how has capitalism come to dominate the global network of economic relationships?

9. Distinguish among core, peripheral, and semiperipheral economies.

10. Explain why Iraq would be considered a peripheral economy.

11. What is odious debt?

12. Which country is considered the strongest and most diverse economy in the world? Why?

13. What contributions do the primary, secondary, and tertiary sectors of the U.S. economy make to the GDP? Which sector contributes the largest share? Explain.

14. What is the difference between a monopoly and an oligopoly?

15. What is a conglomerate? Give an example.

16. Use the example of the banking industry to show how computers and information technologies have shaped the U.S. economy.

17. What does it mean to use computers as informating and automating tools?

18. What do we mean when we say that the United States is an oil- and mineral-dependent economy?

19. Define political system.

20. What is authority? How many types of authority did Weber identify? Give examples of each kind of authority.

21. What are the essential characteristics of a democracy?

22. How do we distinguish between totalitarianism and authoritarian governments?

23. What is a theocracy?

24. What is the power elite? Who makes up the powerelite in the United States?

25. Does C. Wright Mills believe that there are any significant constraints on the decision-making powers of the power elite? Why or why not?

26. Explain the pluralist model of power.

27. What are PACs and 527 groups? Give examples.

28. Define empire, imperialistic power, hegemony, and militaristic power.

29. What are some examples of U.S. power and influence in the world?

30. What are insurgents?

Key Concepts

Economic system	pg. 370
Goods	pg. 370
Services	pg. 370
Domestication	pg. 370
Mechanization	pg. 372
Colonization	pg. 372
Capitalism	pg. 374
Private ownership	pg. 374
Laws of supply and demand	pg. 374
Socialism	pg. 374

Concept Application

Consider the concepts listed below. Match one or more of the concepts with each scenario. Explain your choices.

- ✓ Semiperipheral economy
- ✓ Peripheral economy
- ✓ Conglomerate
- ✓ Primary sector
- ✓ Secondary sector
- ✓ Interest groups

Scenario 1 "In less than three decades, Taiwan has become a major economic player not only in the economy of the Pacific Rim, but in the global system as well. Foreign investors have played a vital role in Taiwan's economic development. For example, a mass buyer like Sears or K-Mart would visit Taiwanese factories and order goods in bulk for sale under the chain's brand name. A company like Arrow shirts or U.S. Shoe would supply samples to several factories and then contract with the factory that offered the best deal in terms of cost and quality. The "Made in Taiwan" label spread worldwide, even if no one outside Taiwan knew a single Taiwanese company that produced the products" (Goldstein 1991).

Scenario 2 At the urging of Chiquita Brands, a unit of the American Financial Corporation and the world's largest banana producer, the Clinton Administration is seeking to overturn an agreement that guarantees small Caribbean banana farmers special access to the European Union market.

"Why is America doing this to us?" Mr. Prosper, 53, asked as his crop was being boxed at a weighing station here the other day. "This is a little place, and this is all that we know, and what we depend on. We have nothing else and we hurt nobody, but now they want to take even this from us."

Much as in neighboring Dominica and St. Vincent and the Grenadines, one-quarter of the labor force in this country of 145,000 people is employed in the banana industry, either growing, processing, or shipping the fruit. In contrast to Central America, where workers paid as little as $2 a day grow most of Chiquita's bananas, Caribbean banana workers are mostly independent growers who own the small plot they farm (Rohter 1997:A6).

Scenario 3 A notable feature of the biggest recent mergers is that the firms dominating this process are not media companies in a strict sense: Disney is avowedly a "family entertainment communication company" in the business of selling theme parks, toys, movies, and videos. Its focus, in the words of CEO Michael Eisner, is on the provision of "non-political entertainment and sports." Time Warner has a similarly wide spectrum of business interests and a comparable marketing orientation. Disney and Time Warner are what Herbert Schiller calls "pop cultural corporate behemoths."

Westinghouse, by contrast, has long been primarily a nuclear power and weapons producer, with its media interests only 10% of sales revenue. With the Westinghouse takeover of CBS, two of the three top networks are controlled by large firms in the politically sensitive nuclear power/weapons industries (the other is NBC's owner General Electric, which along with Westinghouse is one of the top 15 U.S. defense contractors).

Scenario 4 A mainstay of the mining industry is gold, which is being extracted from the West faster than ever before, says France. About 85 percent of the gold extracted in the West ends up in jewelry, the rest going into products such as electronics.

Scenario 5 As more private sector organizations learn to use the tools of the political campaign industry, a broad range of corporations, associations, unions and non-profits are playing a larger, more aggressive role in the shaping of public opinion on matters they deem important.... A study conducted by the Annenberg Public Policy Center of the University of Pennsylvania estimated that during the 1995-96 election cycle one-third of the total dollars spent on advertising in federal elections was attributable to "issue" advocacy efforts.

Applied Research

Find the U.S. Department of Commerce, International Trade Administration website. This site contains information on the 17 countries the U.S. has identified as Big Emerging Markets. Choose 3 or 4 countries and read the corresponding sections on "Economic Trends and Outlooks" and "The Political Environment" for each country. Does there seem to be a connection between the economic outlook and the political environment?

Practice Test: Multiple-Choice Questions

1 _______________ holds the largest amount of proven oil reserves in the world.
   a. Iraq
   b. Saudi Arabia
   c. Kuwait
   d. Afghanistan

2 Which one of the following would be classified as a "service"?
   a. growing food
   b. manufacturing clothing
   c. providing transportation
   d. building computer hardware

3 The most important agricultural revolution in history took place more than 10,000 years ago with the
   a. rise of hunting and gathering societies.
   b. domestication of plants and animals.
   c. invention of the scratch plow.
   d. invention of the wheel.

4 Many of the great Sumerican legacies, such as writing, irrigation, and the wheel are adaptive responses to
   a. the Tigris and Euphrates Rivers.
   b. domestication.
   c. the Nile River.
   d. the industrial revolution.

5 Biblical scholars locate the ____________ region as the setting for Noah's Ark.
   a. Tigris-Euphrates
   b. Sudan
   c. Nile
   d. Indus

6 Under a system of private ownership, _______ own the means of production.
   a. individuals
   b. unions
   c. governments
   d. communes

7 ________ is an essential characteristic of socialist systems.
   a. Public ownership of the means of production
   b. Private ownership of the means of production
   c. The laws of supply are demand
   d. A consumer-driven economy

8 Iraq's economy is heavily dependent on
   a. agriculture.
   b. the service sector.
   c. oil revenues.
   d. migrant labor.

9 According to world system theorists, capitalism has come to dominate the world economy because
   a. under this system, governments control economic activities.
   b. it is the only economic system in the world.
   c. of the ways in which capitalists respond to changes in the economy, especially to economic stagnation.
   d. national interests take precedence over corporate interests.

10 The Vietnamese economy is very vulnerable to price fluxuations. For example, the country managed to become the third largest producer of coffee beans when prices were around $3.00 per pound only to see the price drop to less than 50 cents. The vulnerability explains why Vietnam is classified as a _________ economy.
   a. core
   b. peripheral
   c. semiperipheral
   d. middle-income

11. Iraq's debt began accumulating with the
   a. Gulf War I.
   b. Iran-Iraq War.
   c. Gulf War II.
   d. six-day war.

12 The U.S. economy can be classified as all but which of one of the following?
   a. market oriented
   b. capitalist
   c. socialist
   d. dominated by private enterprise

13. The secondary sector of the economy includes economic activities
   a. that generate or extract raw materials from the natural environment.
   b. that transform raw materials into manufactured goods.
   c. related to delivering services.
   d. related to the creation and distribution of information.

14. Which sector of the economy contributes the most to the GDP of the United States?
   a. primary
   b. secondary
   c. tertiary
   d. manufacturing

15. In the United States, three foreign-owned corporations control 90 percent of technical, medical, and scientific publishing. This situation describes
   a. a monopoly.
   b. an oligopoly.
   c. a conglomerate.
   d. a bureaucracy.

16. Which one of the following factors does <u>not</u> help to explain the drop in union membership?
    a. the increased significance of the manufacturing sector
    b. increased percentage of females in the workforce
    c. increased global competition
    d. increased number of jobs with no union tradition

17. The United States has an estimated 20.2 billion barrels of proven oil reserves. At the current rate of production, these reserves will last about ________________ years.
    a. 10
    b. 20
    c. 40
    d. 100

18. The two top foreign holders of U.S. debt are
    a. Spain and South Africa.
    b. Germany and Italy.
    c. Mexico and Canada.
    d. China and Japan.

19. A "chief," "king," or "queen" possesses power based on which form of authority?
    a. traditional
    b. charismatic
    c. legal-rational
    d. socialistic

20. ________________ leaders often emerge during times of profound crisis.
    a. Traditional
    b. Charismatic
    c. Legal-rational
    d. Socialistic

21. ___________ is a system of government in which power is vested in the citizen body or "the people"
    a. Capitalism
    b. Democracy
    c. Totalitarianism
    d. Authoritarianism

22. The Kirkpatrick doctrine, named after a U.S. Ambassador to the United States in the 1980s, maintains that the U.S. would support ________________ regimes because they are less dangerous to the American way of life.
    a. tertiary
    b. authoritarian
    c. totalitarian
    d. representative

23. _______________ is a form of government in which political authority is in the hands of religious leaders or a theologically-trained elite.
   a. Theocracy
   b. Democracy
   c. Totalitarianism
   d. Authoritarianism

24. C. Wright Mills wrote "The power to make decisions of national and international consequence is now so clearly seated in political, military, and economic institutions that other areas of society seem off to the side." Mills was writing about
   a. monopolies.
   b. the power elite.
   c. a pluralist society.
   d. conglomerates.

25. Jubilee is a special-interest group dedicated to
   a. raising money to support liberal political candidates.
   b. debt relief for the poorest countries in the world.
   c. celebrating diversity.
   d. cleaning up the environment on a global scale.

26. _________________ percent of Iraqi marriages are between first and second cousins.
   a. Seventy
   b. Fifty
   c. Twenty
   d. Five

27. Which "phrase" yields the most hits on the google search engine?
   a. American Empire
   b. British Empire
   c. Roman Empire
   d. Mayan Empire

28. A group of countries under the direct control of a foreign power or government such that the dominant power shapes political, economic, and cultural development is a/an
   a. monopoly.
   b. conglomerate.
   c. PAC.
   d. empire.

29. One U.S Pentagon official pointed out in a news briefing that "the sheer size of this campaign has never been seen before, never been contemplated before." This campaign was
   a. the 1990 oil embargo against Iraq.
   b. the food-for-oil program in Iraq.
   c. the shock and awe campaign in Iraq.
   d. the democratization of Iraq.

True/False Questions

T F 1. The Bush administration is committed to change Iraq's system from a free market to a centrally-planned system.

T F 2. Many agricultural revolutions have taken place over the course of human history.

T F 3. The People's Republic of China, Cuba, and North Korea allow no activities that generate personal wealth.

T F 4. Vietnam fits the profile of a peripheral economy.

T F 5. Luxembourg has the highest per capita GDP in the world.

T F 6. A monopoly exists when a handful of producers dominate a market.

T F 7. Saudi Arabia is the top supplier of petroleum products to the United States.

T F 8. In the United States, Hispanics stand out as the category most likely to vote.

T F 9. P.A.C. stands for Political Action Committee.

T F 10. Israel is the largest buyer of arms in the world.

Internet Resources Related to Economics and Politics

- **Background Notes on the Countries of the World**
  http://www.state.gov/r/pa/ei/bgn/
  The U.S. State Department website contains general information on all countries and covers economic and political arrangements.

Internet Resources Related to Iraq

- **Global Policy Forum; Iraq**
  http://www.globalpolicy.org/security/issues/irqindx.htm
  Global Policy Forum's mission is "to monitor policy making at the United Nations, promote accountability of global decisions, educate and mobilize for global citizen participation, and advocate on vital issues of international peace and justice." Its section on Iraq contains links such Post-War Iraq, Humanitarian Crisis, Oil in Iraq, Weapons Inspection Program, Historical Background and General Articles relating to Iraq.

- **Iraq Daily**
  http://www.iraqdaily.com/
  This website reports daily information on events in Iraq from the World News Network.

- **Iraq Maps**
  http://www.lib.utexas.edu/maps/iraq.html
  This site contains country, city, thematic and detailed maps of Iraq produced by the U.S. Central Intelligence Agency.

Iraq: Background Notes

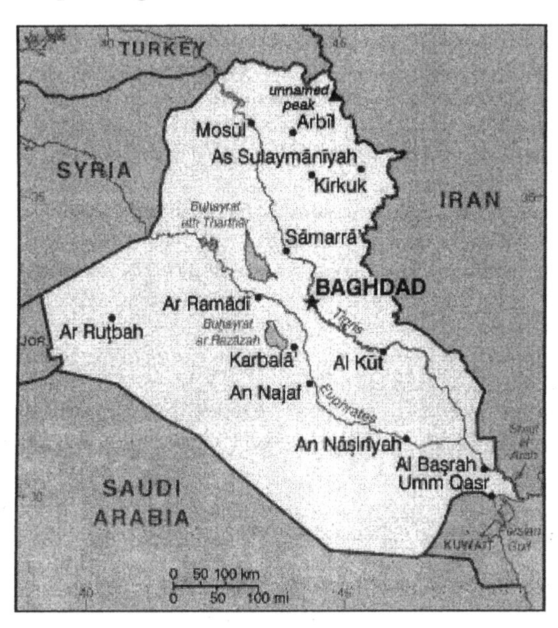

PEOPLE

Nationality: *Noun and adjective*--Iraqi(s).
Population (2002 est.): 24,011,816.
Annual growth rate (2002 est.): 2.82%.
Ethnic groups: Arab 75%-80%, Kurd 15%-20%, Turkman, Chaldean, Assyrian, or others less than 5%.
Religions: Shi'a Muslim 60%, Sunni Muslim 32%-37%, Christian 3%, Yezidi less than 1%.
Languages: Arabic, Kurdish, Assyrian, Armenian.
Education: *Years compulsory*--primary school (age 6 through grade 6). *Literacy*--58%.
Health: *Infant mortality rate* (2002 est.)--57.61 deaths/1,000. *Life expectancy*--67.38 yrs.
Work force (2000, 4.4 million): *Agriculture*--44%; *industry*--26%; *services*--31% (1989 est.).

Almost 75% of Iraq's population live in the flat, alluvial plain stretching southeast toward Baghdad and Basrah to the Persian Gulf. The Tigris and Euphrates Rivers carry about 70 million cubic meters of silt annually to the delta. Known in ancient times as Mesopotamia, the region is the legendary locale of the Garden of Eden. The ruins of Ur, Babylon, and other ancient cities are in Iraq.

Iraq's two largest ethnic groups are Arabs and Kurds. Other distinct groups are Turkomans, Chaldeans, Assyrians, Persians, and Armenians. Arabic is the most commonly spoken language. Kurdish is spoken in the north, and English is the most commonly spoken Western language.

Most Iraqi Muslims are members of the Shi'a sect, but there is a large Sunni population as well, made up of both Arabs and Kurds. Small communities of Christians, Jews, Bahais, Mandaeans, and Yezidis also exist. Most Kurds are Sunni Muslim but differ from their Arab neighbors in language, dress, and customs.

HISTORY

Once known as Mesopotamia, Iraq was the site of flourishing ancient civilizations, including the Sumerian, Babylonian, and Parthian cultures. Muslims conquered Iraq in the seventh century A.D. In the eighth century, the Abassid caliphate established its capital at Baghdad, which became a frontier outpost of the Ottoman Empire.

At the end of World War I, Iraq became a British-mandated territory. When it was declared independent in 1932, the Hashemite family, which also ruled Jordan, ruled as a constitutional monarchy. In 1945, Iraq joined the United Nations and became a founding member of the Arab League. In 1956, the Baghdad Pact allied Iraq, Turkey, Iran, Pakistan, and the United Kingdom, and established its headquarters in Baghdad.

Gen. Abdul Karim Qasim took power in July 1958 coup, during which King Faysal II and Prime Minister Nuri as-Said were killed. Qasim ended Iraq's membership in the Baghdad Pact in 1959. Qasim was assassinated in February 1963, when the Arab Socialist Renaissance Party (Ba'ath Party) took power under the leadership of Gen. Ahmad Hasan al-Bakr as prime minister and Col. Abdul Salam Arif as president.

Nine months later, Arif led a coup ousting the Ba'ath government. In April 1966, Arif was killed in a plane crash and was succeeded by his brother, Gen. Abdul Rahman Mohammad Arif. On July 17, 1968, a group of Ba'athists and military elements overthrew the Arif regime. Ahmad Hasan al-Bakr reemerged as the President of Iraq and Chairman of the Revolutionary Command Council (RCC).

In July 1979, Bakr resigned, and his chosen successor, Saddam Hussein, assumed both offices. The Iran-Iraq war (1980-88) devastated the economy of Iraq. Iraq declared victory in 1988 but actually achieved a weary return to the status quo antebellum. The war left Iraq with the largest military establishment in the Gulf region but with huge debts and an ongoing rebellion by Kurdish elements in the northern mountains. The government suppressed the rebellion by using weapons of mass destruction on civilian targets, including a mass chemical weapons attack on the city of Halabja that killed several thousand civilians.

Iraq invaded Kuwait in August 1990, but a U.S.-led coalition acting under United Nations (UN) resolutions expelled Iraq from Kuwait in February 1991. After the war, the UN Security Council required the regime to surrender its weapons of mass destruction (WMD) and submit to UN inspections. When the Ba'ath regime refused to fully cooperate with the UN inspections, the Security Council employed sanctions to prevent further WMD development and compel Iraqi adherence to international obligations. Coalition forces enforced no-fly zones in southern and northern Iraq to protect Iraqi citizens from attack by the regime and a no-drive zone in southern Iraq to prevent the regime from massing forces to threaten or again invade Kuwait.

A U.S.-led coalition removed the Ba'ath regime in March and April 2003, bringing an end to more than 12 years of Iraqi defiance of UN Security Council resolutions. The coalition, international agencies, and nongovernmental organizations quickly established aid systems, preventing any general humanitarian crisis. The coalition formed the Coalition Provisional Authority (CPA) to provide for the effective administration of Iraq during the period of transitional administration, restore conditions of security and stability, and create conditions in which the Iraqi people can freely determine their own political future. The UN Security Council acknowledged the authorities of the coalition and provided for a role for the UN and other parties to assist in fulfilling these objectives.

The CPA disbanded on June 28, 2004, transferring sovereign authority for governing Iraq to the Iraqi Interim Government (IIG). Based on the timetable laid out in the Law of Administration for the Transitional Period (TAL), the IIG will govern Iraq until a government elected in national elections to be held no later than January 31, 2005 takes office.

ECONOMY

GDP (2003 est.): $38.79 billion.
Annual growth rate (2003 est.): 20%.
GDP per capita (2003 est.): $1,600.
Inflation rate (2003 est.): 27.5%.
Natural resources: Oil, natural gas, phosphates, sulfur.
Agriculture (% of GNP unspecified): *Products*--wheat, barley, rice, vegetables, cotton, dates, cattle, sheep
Industry: (% GNP unspecified): *Types*--petroleum, chemicals, textiles, construction materials, food processing.
Trade: *Exports*--$ 7.542 billion f.o.b (2003 est). *Major markets*--US 37.4%, Taiwan 7.7%, Canada 7.5%, France 7.5%, Jordan 6.9%, Netherlands 5.8%, Italy 4.9%, Morocco 4.3%, Spain 4.1% (2002). *Imports*--$6.521 billion f.o.b (2003 est): food, medicine, manufactures. *Major suppliers*--Jordan 10.4%, France 8.4%, China 7.9%, Vietnam 7.9%, Germany 7.2%, Russia 6.9%, Australia 6.8%, Italy 6.1%, Japan 5.3% (2002).

Historically, Iraq's economy was characterized by a heavy dependence on oil exports and an emphasis on development through central planning. Prior to the outbreak of the war with Iran in September 1980, Iraq's economic prospects were bright. Oil production had reached a level of 3.5 million barrels per day, and oil revenues were $21 billion in 1979 and $27 billion in 1980. At the outbreak of the war, Iraq had amassed an estimated $35 billion in foreign exchange reserves.

The Iran-Iraq war depleted Iraq's foreign exchange reserves, devastated its economy, and left the country saddled with a foreign debt of more than $40 billion. After hostilities ceased, oil exports gradually increased with the construction of new pipelines and the restoration of damaged facilities. Iraq's invasion of Kuwait in August 1990, subsequent international sanctions, damage from military action by an international coalition beginning in January 1991, and neglect of infrastructure drastically reduced economic activity. Government policies of diverting income to key supporters of the regime while sustaining a large military and internal security force further impaired finances, leaving the average Iraqi citizen facing desperate hardships.

Implementation of a UN oil-for-food program in December 1996 improved conditions for the average Iraqi citizen. In December 1999, Iraq was authorized to export unlimited quantities of oil to finance essential civilian needs including, among other things, food, medicine, and infrastructure repair parts. The drop in GDP in 2001-02 was largely the result of the global economic slowdown and lower oil prices. Per capita food imports increased significantly, while medical supplies and health care services steadily improved. The occupation of the US-led coalition in March-April 2003 resulted in the shutdown of much of the central economic administrative structure. The rebuilding of oil, electricity, and other production is proceeding steadily in 2004 with foreign support and despite the continuing internal security incidents. A joint UN and World Bank report released in the fall of 2003 estimated that Iraq's key reconstruction needs through 2007 would cost $55 billion. According to the General Accounting Office as of April 2004, total funds available towards this rebuilding effort include: $21 billion in US appropriations, $18 billion from the Development Fund for Iraq, $2.65 billion in vested and seized assets of the former regime, and $13.6 billion in international pledges. The US and other nations continue assisting Iraqi ministries, to the extent requested by the IIG, and offer extensive economic support.

U.S.-IRAQI RELATIONS

As the lead nation in the international coalition which removed the Ba'ath regime, the United States is committed to the establishment of a stable, united, prosperous, and democratic Iraq. U.S. forces remain in Iraq as part of the Multi-National Force-Iraq to assist the IIG to train its security forces, as well as to work in partnership with the IIG to combat forces that seek to derail Iraq's progression to full democracy. The U.S. Government is carrying out a multibillion-dollar program to assist in the reconstruction of Iraq.

Sources: U.S. Department of State (2004)
http://www.state.gov

Chapter References

DiSilvestro, Roger. 1996. "Investigating the Last Great American Gold Heist." *National Wildlife*. (Dec-Jan v35 n1 p70(1)

Faucheux, Ron. 1998. "The Indirect Approach." *Campaigns & Elections*. (June):v19 n6 p18(6)

Goldstein, Steven M. 1991. *Minidragons: Fragile Economic Miracles in the Pacific*. New York: Ambrose Video.

Herman, Edward S. 1996. "The Media Mega-Mergers." *Dollars & Sense*. (May-June):n205 p8(6)

Rohter, Larry. 1997. "Trade Storm Imperils Caribbean Banana Crops." *The New York Times* (May 9):A6.

Answers

Concept Application

1.	Semiperipheral economy	pg. 377
2.	Peripheral economy	pg. 377
3.	Conglomerate	pg. 381
4.	Primary Sector; Secondary sector	pg. 380; 380
5.	Special Interest Groups	pg. 398

Multiple-Choice

1.	b	pg. 369
2.	c	pg. 370
3.	b	pg. 370
4.	a	pg. 370
5.	a	pg. 370
6.	a	pg. 374
7.	a	pg. 375
8.	c	pg. 374
9.	c	pg. 375
10.	b	pg. 377
11.	b	pg. 380
12.	c	pg. 380
13.	b	pg. 380
14.	c	pg. 381
15.	b	pg. 381
16.	a	pg. 384
17.	a	pg. 385
18.	d	pg. 385
19.	a	pg. 387
20.	b	pg. 388
21.	b	pg. 388
22.	b	pg. 391
23.	a	pg. 392
24.	b	pg. 395
25.	b	pg. 397
26	b	pg. 398
27.	a	pg. 399
28.	d	pg. 399
29.	c	pg. 400

True/False

1.	F	pg. 368
2.	T	pg. 370
3.	F	pg. 374
4.	T	pg. 377
5.	T	pg. 380
6.	F	pg. 381
7.	T	pg. 386
8.	F	pg. 389
9.	T	pg. 398
10.	F	pg. 400

Chapter 12

Family and Aging
With Emphasis on Japan

Study Questions

1. Why is Japan the focus of a chapter on family and ~~aging~~? How does the U.S. compare to Japan on indicators related to family well-being and stability?

Japan has low infant mortality, divorce rates, teen birth rates, single parent. Elders live with families
US has higher marriage rate.

2. Why is "family" a difficult concept to define? What are some criteria that might be used to define family?

so many different views - some recognize kinship, membership characteristics, legal recognition and or social function.

3. How does the family contribute to order and stability in society? What are some problems with defining family in terms of social functions?

Regulate sexual behaviour, replace dead members, socialize the young, provide care/support, cofer social status.
- always exceptions.

4. How would a conflict theorist respond to a functionalist's assessment of the family?

- may be competing interests, actions might not benefit everyone,

5. Distinguish between productive and reproductive work. Which type of work is more valued?

* productive = ~~existence~~, food, clothing, shelter
reproductive = child bearing, caregiving, educating managing household

6. How is family related to social inequality in society?

Social status - power, wealth, property, privalages differs depending on if you are a women or man, black or white.

7. How has family created racial divisions and boundaries?

exogamy - encourage marrying someone of a different social category

endogamy - marrying within the same social category.

can be formal/informal norms of society

8. Describe at least three major changes in American family life since 1900.

- 90s - 2% families where both parents worked outside of home
- 2000 - 44% families where both parents work
- life expectancy increased 28/30 yrs
- infant mortality dropped from 99 to 6 deaths per 1000

9. How did the Industrial Revolution destroy the household-based economy and lead to the breadwinner system?

Before industrialization workplace was home/farm
- men hunted - women cooked
- industrialization split the workplace - took production from women = breadwinner system

10. According to Kingsley Davis, what strains and demographic factors led to the collapse of the breadwinner system?

- too much strain - men kept from home - women weren't important in production
- strain on men as sole producer for the family
- decreased fertility rates, ↑ life expectancy, ↑ divorce, ↑ opportunities for work suitable to women.

11. Describe at least three major changes in Japanese family life since 1900.

- sharp decline in infant mortality
- low fertility rate
- increased life expectancy

12. What caused the *ie* family system to fall? What system replaced it? What belief underlies the connection between family and Japan's welfare system?

When the US occupied Japan after WWII they ended the ie. was replaced by the nuclear family. Breadwinner system takes over.

13. Explain: "Japan does not have a couple's culture."

- used to Arranged marriages
- don't have places to "date"
- television doesn't pressure marriage
- girls do their thing, guys do theirs.

14. What is a "parasite single." Explain the "new single concept."

Parasite single - singles who live with their parents but make no cost contributions.

new single concept - living a comfortable life with no financial/emotional pressures of parenting.

15. How is Japan's employment system connected to the country's low fertility rate?

- less children - more chances for women to work
- children grow up = go back to work

16. In general, how do economic arrangements shape the character of sexual stratification?

- peope use economic, political and physical sources to dominate
- males are the "ideal" = stratification
- specialized work

17. What is intimacy at a distance? What factors gave rise to this phenomenon?

the distance between adult-child relationships
- due to employing females, no elders in the house, kids going to school, age-segregation.

18. How has the status of children been affected by industrialization?

- children lost their economic value
- provide emotional services

19. How do increases in life expectancy alter the composition of the family?

- more familytime - parents live longer
- marriage length increases
- more time to marry, settle, get a job

20. What is "caregiver burden"? Is care giving only a burden?

The extent to which caregivers believe that their emotional balance, physical heath, social life and financial status suffer due to their role.

21. What are some of the major differences between the elderly-caregiver relationship in Japan and the U.S.?

US - seniors live active lives
- live alone or with spouce
- have caregivers if needed

Japan - live with their children

Key Concepts

Aging Population	pg. 408
Fertility rate	pg. 408
Caregiver burden	pg. 438
Family Type	
Ideal	pg. 409
Nuclear	pg. 412
Extended	pg. 412
Single-parent	pg. 412
Household	pg. 412
Life Chances	pg. 413
Secure parental employment	pg. 416
Productive work	pg. 415
Reproductive work	pg. 415
Choice of Spouses	
Arranged	pg. 412
Romantic	pg. 412
Endogamy	pg. 417
Exogamy	pg. 417
Homogamy	pg. 412
Family Authority	
Patriarchal	pg. 412
Matriarchal	pg. 412
Equilitarian	pg. 412

p 412

Family Residence
- Patrilocal pg. 412
- Matrilocal pg. 412
- Neolocal pg. 412

Number of Marriage Partners
- Monogamy pg. 412
- Polygamy pg. 412
  - Polygyny pg. 412
  - Polyandry pg. 412

Descent
- Patrilineal pg. 412
- Matrilineal pg. 412
- Bilateral pg. 412

Household Types
- Low-technology tribal societies pg. 430
- Surplus wealth pg. 430
- Fortified households pg. 433
- Private households pg. 434
- Advanced market economies pg. 434

Concept Application

Consider the concepts listed below. Match one or more of the concepts with each scenario. Explain your choices.

- ✓ Aging populations
- ✓ Caregiver burden
- ✓ Exogamy
- ✓ Fertility rate
- ✓ Reproductive work

Scenario 1 Sam works hard at his job in the factory. His supervisor knows him as a diligent, focused employee. He barely missed a day of work during his first five years on the job. A year ago, however, Sam's mother was diagnosed with Alzheimer's disease, and he decided to have her cared for in his home. But costs mounted quickly, and after a few months, Sam could only afford to have the home health care worker visit three days a week. Now Sam is struggling to balance his job responsibilities with caring for his mother. He's almost exhausted his supply of sick days, his lack of concentration at work has caused some costly mistakes, and his supervisor's patience is at an end. Something's got to give (Guttchen and Pettigrew 2000, p. 31).

Scenario 2 Recently, the largest-circulation Jewish newspaper in the country carried an opinion article pronouncing, with equanimity, that "the Jewish taboo on mixed marriage has clearly collapsed." Around the same time, and more startlingly, the *New York Times* published a photograph taken at the nuptials of a male rabbi and a female Protestant minister, a rite that was itself blessed by an assemblage of priests, ministers, and rabbis, all standing together under a Jewish wedding canopy. What this powerfully suggestive photograph tells us is not just that many American Jews, including at least some of the rabbis among them, have abandoned long-standing communal norms but that they, again including at least some of the rabbis among them, seem to have replaced those norms with an entirely new set of beliefs about what constitutes an authentic expression of Judaism--and what, if anything, lies beyond the limits of such expression. Long in the building, the intermarriage crisis is now propelling a massive transformation of American Jewish life.

Scenario 3 Next year, for the first time in history, people over 60 will outnumber kids 14 or younger in industrial countries. Even more startling, the population of the Third World, while still comparatively youthful, is aging faster than that of the rest of the world. In France, for example, it took 140 years for the proportion of the population age 65 or older to double from 9 percent to 18 percent. In China, the same feat will take just 34 years. In Venezuela, 22 (Longman 1999, p. 30).

Scenario 4 Italians have stopped making babies; the nation is ageing fast; and, according to the country's chief statistical body, [Italian] women now bear 1.2 babies apiece. Only the Spaniards, in Western Europe, are as unproductive. At last count, in 1996, deaths had outpaced births for four years in a row. If Italy's population is slightly up, it is thanks to the 178,000 immigrants who took up legal residence two years ago (The Economist 1998, p. 51).

Scenario 5 Whether at home or in the hospital, sons and daughters play a crucial role in medical treatment and care [in China]. A scarcity of medical resources, characteristic of developing economies, forces hospitals to rely on the work of family members who provide food, purchase and administer medicine, deliver and pick up lab tests and x-rays, and monitor and bathe the patients. Relatives draw on their personal connections to doctors and nurses to obtain treatment and hospital beds (Otis 2001, p. 471).

Applied Research

Put together a family tree tracing the maternal and paternal side of the family. Interview parents and grandparents. How far can each party go back in time?

Practice Test: Multiple-Choice Questions

1. Japan's _____________ is a major national concern.
   a. high death rate
   b. low dependency ratio
   c. low fertility rate
   d. high infant mortality rate

2. In comparison to the United States, Japan has a
   a. higher total fertility rate.
   b. lower rate of reported domestic abuse cases.
   c. higher divorce rate.
   d. lower life expectancy.

3. An aging population is a label attached to a situation in which
   a. the number of elderly is increasing in a society.
   b. one out of every three people is 65 and over.
   c. the youth outnumber the elderly population.
   d. the percentage of the population age 65 and older is increasing relative to other age groups.

4. Which one of the following constitutes primary kin?
   a. mother, father, sister, brother
   b. mother's mother, mother's father, sister's son
   c. brother's daughter's son
   d. brother's daughter's son's son

5. People make decisions about which kin they will acknowledge as family and which kin they will "forget." This process is known as
   a. amnesia and recall.
   b. self-fulfilling prophecy.
   c. selective remembering and forgetting.
   d. differential association.

6. Sara marries someone of the same religion as herself. She has followed the norm of
   a. patrilocal.
   b. endogamy.
   c. exogamy.
   d. monogamy.

7. In at least 65 countries in the world the fertility rate is below ____________, the rate needed to replace those members who die.
   a. 1.8
   b. 2.1
   c. 3.2
   d. 4.0

8. __________ is the author of *The Origin of Family, Private Property and the State.*
   a. Frederich Engles
   b. Karl Marx
   c. Emile Durkheim
   d. Max Weber

9. Which one of the following countries is the most likely to have one of the lowest fertility rates in the world?
   a. Uganda
   b. Afghanistan
   c. United States
   d. Canada

10. Child bearing, caregiving, managing households, and educating children fall under the category of
    a. productive work.
    b. reproductive work.
    c. life chances.
    d. housework.

11. Ji-wu lives in a household where his father is unemployed but his mother works 35 hours per week at a job she has held for five years. According to the U.S. Census Bureau, Ji-wu lives in a household with
    a. a dead-beat dad.
    b. poverty-level income.
    c. insecure parental employment.
    d. secure parental employment.

12. At one time (before 1967) the United States had laws prohibiting marriages between people classified as white and black. Those laws enforced
    a. polyandry.
    b. monogamy.
    c. endogamy.
    d. exogamy.

13. Currently, Japan issues 50,000 work visas per year to foreigners. Demographers project this number must increase to ___________ per year to prevent its population from shrinking.
    a. 5 million
    b. 1.2 million
    c. 640,000
    d. 100,000

14. The breadwinner system that Davis described did not last because it placed too much strain on husbands and wives. The strain stemmed from all but one of the following sources:
    a. never before had the roles of husband and wife been so distant.
    b. never before had women played such an indirect role in producing what the family consumed.
    c. never before had men had it so easy relative to the role of women.
    d. never before had men had to bear the sole responsibility of supporting the family.

15. Kingsley Davis believed that married women became motivated to seek work outside the household because of all but which one of the following reasons?
    a. changes in child bearing experiences
    b. boredom with children and housework
    c. increases in life expectancy
    d. rising divorce rates

16. The *ie* family system in Japan was abolished during what time period?
    a. after WWI
    b. in 1898 with the rise of the Domestic Relations and Inheritance Laws
    c. after WWII when the U.S. occupied Japan
    d. during the economic crisis of the 1990s

17. In Japan, the population of working single adults (22 and older) who live with their parents while contributing little to household expenses are known as
    a. the baby boomlet.
    b. spoiled singles.
    c. parasite singles.
    d. mama's boys and girls.

18. The *juku* pressure is stressful for everyone involved, but especially for
    a. children.
    b. fathers.
    c. grandparents.
    d. mothers.

19. In Japan, junior colleges train young women to be
    a. skilled, gracious, and responsible homemakers.
    b. entry-level employees.
    c. secretaries.
    d. eldercare providers.

20. Sociologist Randall Collins argues that women must __________ if they are to be men's equals.
    a. be valued as mothers
    b. have access to combat roles in the military
    c. have access to agents of violence control
    d. become involved in athletics at an early age

21. _____________ emerge with the establishment of a market economy, a centralized bureaucratic state, and agencies of social control.
   a. Low-technology tribal societies
   b. Fortified households
   c. Private households
   d. Advanced market economies

22. The case of Chico Mendes was used to illustrate
   a. the status of wealthy children who happen to live in labor-intensive environments.
   b. the role of children as consumers.
   c. the plight of children in Brazil.
   d. the economic role of children in labor-intensive poor countries.

23. Increases in life expectancy have altered family life in all but which one of the following ways?
   a. the chances that children will lose one or both parents before age 16 has decreased dramatically
   b. the percentages of elderly people living in nursing homes has increased dramatically
   c. the number of people surviving to old age has increased
   d. people have time to choose and get to know a partner

24. Someone praises a Japanese mother whose son earned a grade of 100% on a math exam by saying "He is very smart, isn't he?" Which one of the following represents her likely response?
   a. I know, he studied so hard.
   b. No, he is not so smart. He was just lucky.
   c. Yes, he is just naturally good at math.
   d. I don't know how he got to be so smart.

True/False Questions

T F 1. The size of any given person's family network is beyond calculation (i.e., comprehension).

T F 2. An aging population is one in which the percentage of the population 65 and over is increasing relative to other age groups.

T F 3. Federal law defines a spouse as a person of the opposite sex who is a husband or wife.

T F 4. It appears that American males do more housework than their counterparts in Japan.

T F 5. In the U.S. children classified as Asian are least likely to live in secure parental employment households.

T F 6. There are no arranged marriages in Japan today.

T F 7. For the most part, Japanese women are expected to quit working when they marry or have children.

T F 8. In Japan, approximately 50 percent of elderly persons reside in nursing homes.

T F 9. Approximately 30 percent of American elderly live in nursing homes.

Internet Resources Related to Family and Aging

- **The Future of Children**
  http://www.futureofchildren.org/
  The Future of Children is an on-line journal published three times a year by the Center for the Future of Children. The journal's primary purpose is to "disseminate timely information on major issues related to children's well-being, with special emphasis on providing objective analysis and evaluation, translating existing knowledge into effective programs and policies, and promoting constructive institutional change." Articles relating to children's health and development can also be accessed through this site. Each journal issue examines a specific topic such as low-birth weight babies, juvenile court, special education for children, and early childhood programs.

- **American Association of Retired Persons [AARP]**
  http://www.aarp.org
  AARP is a "nonprofit, nonpartisan membership organization for people over 50." The research link offers data and analysis of issues important to this age group.

Internet Resources Related to Japan

- ***Japan Times* Online**
  http://www.japantimes.co.jp
  For the latest news, opinions, and coverage of social and cultural events check on *Japanese Times* Online.

- **Web Japan Gateway for All Japanese Information**
  http://web-japan.org/stat/
  Search crime, education, economy, medical care, opinion surveys, leisure, politics, and other links for information on all things Japanese.

Japan: Background Notes

PEOPLE

Nationality: *Noun and adjective*--Japanese.
Population (2004 est.): 127.7 million.
Population growth rate (2003 est.): 0.11%.
Ethnic groups: Japanese; Korean (0.6%).
Religions: Shinto and Buddhist; Christian (about 0.7%).
Language: Japanese.
Education: *Literacy*--99%.
Health (2003): *Infant mortality rate*--3.3/1,000. *Life expectancy*--males 77 yrs., females 84 yrs.
Work force (67 million, 2003): *services*--42%; *trade, manufacturing, mining, and construction*-- 46%; *agriculture, forestry, fisheries*-- 5%; *government*--3%.

Japan's population, currently some 128 million, has experienced a phenomenal growth rate during the past 100 years as a result of scientific, industrial, and sociological changes, but this has recently slowed because of falling birth rates. High sanitary and health standards produce a life expectancy exceeding that of the United States.

Japan is an urban society with only about 6% of the labor force engaged in agriculture. Many farmers supplement their income with part-time jobs in nearby towns and cities. About 80 million of the urban population is heavily concentrated on the Pacific shore of Honshu and in northern Kyushu. Major population centers include: Metropolitan Tokyo with approximately 14 million; Yokohama with 3.3 million; Osaka with 2.6 million; Nagoya with 2.1 million; Kyoto with 1.5 million; Sapporo with 1.6 million; Kobe with 1.4 million; and Kitakyushu, Kawasaki, and Fukuoka with 1.2 million each. Japan faces the same problems that confront urban industrialized societies throughout the world: overcrowded cities, congested highways, air pollution, and rising juvenile delinquency.

Shintoism and Buddhism are Japan's two principal religions. Shintoism is founded on myths and legends emanating from the early animistic worship of natural phenomena. Since it was unconcerned with problems of afterlife which dominate Buddhist thought, and since Buddhism easily accommodated itself to local faiths, the two religions comfortably coexisted, and Shinto shrines and Buddhist monasteries often became administratively linked. Today many Japanese are adherents of both faiths. From the 16th to the 19th century Shintoism flourished.

Adopted by the leaders of the Meiji restoration, it received state support and was cultivated as a spur to patriotic and nationalistic feelings. Following World War II, state support was discontinued, and the emperor disavowed divinity. Today Shintoism plays a more peripheral role in the life of the Japanese people. The numerous shrines are visited regularly by a few believers and, if they are historically famous or known for natural beauty, by many sightseers. Many marriages are held in the shrines, and children are brought there after birth and on certain anniversary dates; special shrine days are celebrated for certain occasions, and numerous festivals are held throughout the year. Many homes have "god shelves" where offerings can be made to Shinto deities.

Buddhism first came to Japan in the 6th century and for the next 10 centuries exerted profound influence on its intellectual, artistic, social, and political life. Most funerals are conducted by Buddhist priests, and burial grounds attached to temples are used by both Buddhist and Shinto faiths.

Confucianism arrived with the first great wave of Chinese influence into Japan between the 6th and 9th centuries. Overshadowed by Buddhism, it survived as an organized philosophy into the late 19th century and remains today as an important influence on Japanese thought and values.

Christianity, first introduced into Japan in 1549, was virtually stamped out by the government a century later; it was reintroduced in the late 1800s and has spread slowly. Today it has 1.4 million adherents, including a relatively high percentage of important figures in education and public affairs.

Beyond the three traditional religions, many Japanese today are turning to a great variety of popular religious movements normally lumped together under the name "new religions." These religions draw on the concept of Shinto, Buddhism, and folk superstition and have developed in part to meet the social needs of elements of the population. The officially recognized new religions number in the hundreds, and total membership is reportedly in the tens of millions.

HISTORY

Japanese legend maintains that Japan was founded in 600 BC by the Emperor Jimmu, a direct descendant of the sun goddess and ancestor of the present ruling imperial family. About AD 405, the Japanese court officially adopted the Chinese writing system. Together with the introduction of Buddhism in the sixth century, these two events revolutionized Japanese culture and marked the beginning of a long period of Chinese cultural influence. From the establishment of the first fixed capital at Nara in 710 until 1867, the emperors of the Yamato dynasty were the nominal rulers, but actual power was usually held by powerful court nobles, regents, or "shoguns" (military governors).

Contact With the West

The first recorded contact with the West occurred about 1542, when a Portuguese ship, blown off its course to China, landed in Japan. During the next century, traders from Portugal, the Netherlands, England, and Spain arrived, as did Jesuit, Dominican, and Franciscan missionaries. During the early part of the 17th century, Japan's shogunate suspected that the traders and missionaries were actually forerunners of a military conquest by European powers. This caused the shogunate to place foreigners under progressively tighter restrictions. Ultimately, Japan forced all foreigners to leave and barred all relations with the outside world except for severely restricted commercial contacts with Dutch and Chinese merchants at Nagasaki. This isolation lasted for 200 years, until Commodore Matthew Perry of the U.S. Navy achieved the opening of Japan to the West with the Convention of Kanagawa in 1854.

Within several years, renewed contact with the West profoundly altered Japanese society. The shogunate resigned, and the emperor was restored to power. The "Meiji restoration" of 1868 initiated many reforms. The feudal system was abolished, and numerous Western institutions were adopted, including a Western legal system and constitutional government along quasi-parliamentary lines.

In 1898, the last of the "unequal treaties" with Western powers was removed, signaling Japan's new status among the nations of the world. In a few decades, by creating modern social, educational, economic, military, and industrial systems, the Emperor Meiji's "controlled revolution" had transformed a feudal and isolated state into a world power.

Wars With China and Russia

Japanese leaders of the late 19th century regarded the Korean Peninsula as a "dagger pointed at the heart of Japan." It was over Korea that Japan became involved in war with the Chinese Empire in 1894-95 and with Russia in 1904-05. The war with China established Japan's domination of Korea, while also giving it the Pescadores Islands and Formosa (now Taiwan). After Japan defeated Russia in 1905, the resulting Treaty of Portsmouth awarded Japan certain rights in Manchuria and in southern Sakhalin, which Russia had received in 1875 in exchange for the Kurile Islands. Both wars gave Japan a free hand in Korea, which it formally annexed in 1910.

World War I to 1952

World War I permitted Japan, which fought on the side of the victorious Allies, to expand its influence in Asia and its territorial holdings in the Pacific. The postwar era brought Japan unprecedented prosperity. Japan went to the peace conference at Versailles in 1919 as one of the great military and industrial powers of the world and received official recognition as one of the "Big Five" of the new international order. It joined the League of Nations and received a mandate over Pacific islands north of the Equator formerly held by Germany.

During the 1920s, Japan progressed toward a democratic system of government. However, parliamentary government was not rooted deeply enough to withstand the economic and political pressures of the 1930s, during which military leaders became increasingly influential.

Japan invaded Manchuria in 1931 and set up the puppet state of Manchukuo. In 1933, Japan resigned from the League of Nations. The Japanese invasion of China in 1937 followed Japan's signing of the "anti-Comintern pact" with Nazi Germany the previous year and was part of a chain of developments culminating in the Japanese attack on the United States at Pearl Harbor, Hawaii, on December 7, 1941.

After almost 4 years of war, resulting in the loss of 3 million Japanese lives and the atomic bombings of Hiroshima and Nagasaki, Japan signed an instrument of surrender on the U.S.S. *Missouri* in Tokyo Harbor on September 2, 1945. As a result of World War II, Japan lost all of its overseas possessions and retained only the home islands. Manchukuo was dissolved, and Manchuria was returned to China; Japan renounced all claims to Formosa; Korea was occupied and divided by the U.S. and the U.S.S.R.; southern Sakhalin and the Kuriles were occupied by the U.S.S.R.; and the U.S. became the sole administering authority of the Ryukyu, Bonin, and Volcano Islands. The 1972 reversion of Okinawa completed the U.S. return of control of these islands to Japan.

After the war, Japan was placed under international control of the Allies through the Supreme Commander, Gen. Douglas MacArthur. U.S. objectives were to ensure that Japan would become a peaceful nation and to establish democratic self-government supported by the freely expressed will of the people. Political, economic, and social reforms were introduced, such as a freely elected Japanese Diet (legislature) and universal adult suffrage. The country's constitution took effect on May 3, 1947. The United States and 45 other Allied nations signed the Treaty of Peace with Japan in September 1951. The U.S. Senate ratified the treaty in March 1952, and under the terms of the treaty, Japan regained full sovereignty on April 28, 1952.

ECONOMY

Japan's industrialized, free market economy is the second-largest in the world. Its economy is highly efficient and competitive in areas linked to international trade, but productivity is far lower in areas such as agriculture, distribution, and services. After achieving one of the highest economic growth rates in the world from the 1960s through the 1980s, the Japanese economy slowed dramatically in the early 1990s, when the "bubble economy" collapsed.

Japan's reservoir of industrial leadership and technicians, well-educated and industrious work force, high savings and investment rates, and intensive promotion of industrial development and foreign trade have produced a mature industrial economy. Japan has few natural resources, and trade helps it earn the foreign exchange needed to purchase raw materials for its economy.

While Japan's long-term economic prospects are considered good, Japan is currently in its worst period of economic growth since World War II. Plummeting stock and real estate prices in the early 1990s marked the end of the "bubble economy." The impact of the Asian financial crisis of 1997-98 also was substantial. Real GDP in Japan grew at an average of roughly 1% yearly in the 1990s, compared to growth in the 1980s of about 4% per year. Real growth in 2003 was 2.7%.

Agriculture, Energy, and Minerals

Only 15% of Japan's land is suitable for cultivation. The agricultural economy is highly subsidized and protected. With per hectare crop yields among the highest in the world, Japan maintains an overall agricultural self-sufficiency rate of about 50% on fewer than 5.6 million cultivated hectares (14 million acres). Japan normally produces a slight surplus of rice but imports large quantities of wheat, sorghum, and soybeans, primarily from the United States. Japan is the largest market for U.S. agricultural exports.

Given its heavy dependence on imported energy, Japan has aimed to diversify its sources. Since the oil shocks of the 1970s, Japan has reduced dependence on petroleum as a source of energy from more than 75% in 1973 to about 57% at present. Other important energy sources are coal, liquefied natural gas, nuclear power, and hydropower.

Deposits of gold, magnesium, and silver meet current industrial demands, but Japan is dependent on foreign sources for many of the minerals essential to modern industry. Iron ore, coke, copper, and bauxite must be imported, as must many forest products.

Labor

Japan's labor force consists of some 67 million workers, 40% of whom are women. Labor union membership is about 12 million. The unemployment rate is currently around 5%, still near the post-war high. In 1989, the predominantly public sector union confederation, SOHYO (General Council of Trade Unions of Japan), merged with RENGO (Japanese Private Sector Trade Union Confederation) to form the Japanese Trade Union Confederation.

FOREIGN RELATIONS

Despite its current slow economic growth, Japan remains a major economic power both in Asia and globally. Japan has diplomatic relations with nearly all independent nations and has been an active member of the United Nations since 1956. Japanese foreign policy has aimed to promote peace and prosperity for the Japanese people by working closely with the West and supporting the United Nations.

In recent years, the Japanese public has shown a substantially greater awareness of security issues and increasing support for the Self Defense Forces. This is in part due to the Self Defense Forces' success in disaster relief efforts at home, and its participation in peacekeeping operations such as in Cambodia in the early 1990s. However, there are still significant political and psychological constraints on strengthening Japan's security profile. Although a military role for Japan in international affairs is highly constrained by its constitution and government policy, Japanese cooperation with the United States through the 1960 U.S.-Japan Security Treaty has been important to the peace and stability of East Asia. Currently, there are domestic discussions about possible reinterpretation or revision of Article 9 of the Japanese constitution. All postwar Japanese governments have relied on a close relationship with the United States as the foundation of their foreign policy and have depended on the Mutual Security Treaty for strategic protection.

While maintaining its relationship with the United States, Japan has diversified and expanded its ties with other nations. Good relations with its neighbors continue to be of vital interest. After the signing of a peace and friendship treaty with China in 1978, ties between the two countries developed rapidly. Japan extended significant economic assistance to the Chinese in various modernization projects and supported Chinese membership in the World Trade Organization (WTO). Japan's economic assistance to China is now declining. The development of political relations is hampered by China's opposition to Prime Minister Koizumi's visits to the Yasukuni Shrine war memorial and historical and territorial issues. At the same time, Japan maintains economic but not diplomatic relations with Taiwan, with which a strong bilateral trade relationship thrives.

Japan's ties with South Korea have improved since an exchange of visits in the mid-1980s by their political leaders. Japan has limited economic and commercial ties with North Korea. A surprise visit by Prime Minister Koizumi to Pyongyang on September 17, 2002, resulted in renewed discussions on contentious bilateral issues--especially that of abductions to North Korea of Japanese citizens--and Japan's agreement to resume normalization talks in the near future. In October 2002, five abductees returned to Japan, but soon after negotiations reached a stalemate over the fate of abductees' families in North Korea. Japan strongly supported the United States in its efforts to encourage Pyongyang to abide by the nuclear Non-Proliferation Treaty and its agreements with the International Atomic Energy Agency (IAEA). Japan is continuing to cooperate with the U.S. in international efforts to get Pyongyang to abandon development of weapons of mass destruction. The U.S., Japan, and South Korea closely coordinate and consult trilaterally on policy toward North Korea, and Japan participates in the Six-Party talks to end North Korea's nuclear arms ambitions.

Japan's relations with Russia are hampered by the two sides' inability to resolve their territorial dispute over the islands that make up the Northern Territories (Kuriles) seized by the U.S.S.R. at the end of World War II. The stalemate has prevented conclusion of a peace treaty formally ending the war between Japan and Russia. The United States supports Japan on the Northern Territories issue and recognizes Japanese sovereignty over the islands. Despite the lack of progress in resolving the Northern Territories dispute, however, Japan and Russia have made progress in developing other aspects of the relationship.

Beyond relations with its immediate neighbors, Japan has pursued a more active foreign policy in recent years, recognizing the responsibility that accompanies its economic strength. It has expanded ties with the Middle East, which provides most of its oil, and has been the second-largest assistance donor (behind the U.S.) to Iraq and Afghanistan. Japan increasingly is active in Africa and Latin America–recently concluding negotiations with Mexico on an Economic Partnership Agreement (EPA)--and has extended significant support to development projects in both regions. A Japanese-conceived peace plan became the foundation for nationwide elections in Cambodia in 1998. Japan's economic engagement with its neighbors is increasing, as evidenced by the conclusion of an EPA with Singapore, and its ongoing negotiations for EPAs with Korea, Thailand, Malaysia, and the Philippines.

U.S.-JAPAN RELATIONS

The U.S.-Japan alliance is the cornerstone of U.S. security interests in Asia and is fundamental to regional stability and prosperity. Despite the changes in the post-Cold War strategic landscape, the U.S.-Japan alliance continues to be based on shared vital interests and values. These include stability in the Asia-Pacific region, the preservation and promotion of political and economic freedoms, support for human rights and democratic institutions, and securing of prosperity for the people of both countries and the international community as a whole.

Japan provides bases and financial and material support to U.S. forward-deployed forces, which are essential for maintaining stability in the region. Under the U.S.-Japan Treaty of Mutual Cooperation and Security, Japan hosts a carrier battle group, the III Marine Expeditionary Force, the 5th Air Force, and the Army's 9th Theater Support Command. The United States currently maintains approximately 53,000 troops in Japan, about half of whom are stationed in Okinawa.

Over the past several years the alliance has been strengthened through revised Defense Guidelines, which expand Japan's noncombat role in a regional contingency. The alliance has also been strengthened by the Special Action Committee on Okinawa (SACO) program to consolidate U.S. military presence in Okinawa, the 2001 5-year agreement on Host Nation Support of U.S. forces stationed in Japan, and technical cooperation on ballistic missile defense. After the tragic events of September 11, 2001, Japan has participated significantly with the global war on terrorism by providing major logistical support for U.S. and coalition forces in the Indian Ocean. Japan also has played a leading role in the reconstruction of Afghanistan, as well as in the political and economic rehabilitation of Iraq. Their efforts include the passage of historic legislation allowing Japan's Self Defense Forces to participate in reconstruction and humanitarian missions in Iraq; by April 2004, nearly 1,000 Self Defense Force troops were operating in the southern Iraqi city of Al Samawah.

Source: Department of State (2004)
http://www.state.gov

Chapter References

The Economist. 1998. "Why Italians don't make babies." 347(8067):53.

Guttchen, David, and Mary L. Pettigrew. 2000. "Easing the Caregiver Burden." *Risk & Insurance* 11(11):31.

Longman, Phillip J. 1999. "The World Turns Gray." *U.S. News & World Report* 126(8):30.

Otis, Eileen M. 2001. Review of *Giving Care, Writing Self: A "New" Ethnography* by J. Schneider and W. Laihua. New York: Peter Lang.

Wertheimer, Jack. 2001. "Surrendering to Intermarriage." Commentary 111(3):25

Answers

Concept Application

1. Caregiver burden	pg. 408
2. Exogamy	pg. 438
3. Aging populations	pg. 417
4. Fertility rate	pg. 408
5. Reproductive Work	pg. 415

Multiple-Choice

1.	c	pg. 407	13.	c	pg. 418
2.	b	pg. 410	14.	c	pg. 421
3.	d	pg. 408	15.	b	pg. 421+
4.	a	pg. 409	16.	c	pg. 427
5.	c	pg. 409	17.	c	pg. 428
6.	b	pg. 413	18.	d	pg. 428
7.	b	pg. 413	19.	a	pg. 429
8.	a	pg. 414	20.	c	pg. 430
9.	d	pg. 414	21.	c	pg. 434
10.	b	pg. 415	22.	d	pg. 436
11.	d	pg. 416	23.	b	pg. 437+
12.	c	pg. 417	24.	b	pg. 437

True/False

1.	T	pg. 408
2.	T	pg. 407
3.	T	pg. 410
4.	T	pg. 415
5.	F	pg. 417
6.	F	pg. 428
7.	T	pg. 429
8.	F	pg. 440
9.	F	pg. 439

Chapter 13

Population and Urbanization
With Emphasis on India

Study Questions

1. Why is India the focus of a chapter on population? How does the U.S. compare to India in terms of population size?

2nd largest pop – gunna be #1
India has 784 mill more people than US (1/3 the size)
↳ 17% of pop, 2.4% landmass

2. How is population size determined? Explain.

3. What is a population pyramid? What shapes can it take?

bar graph representing age/sex composition of a pop.
– expansive – triangular, pop is increasing, lots of young people. (India)
– constrictive – narrower base than middle, lots of middle aged (US)
– stationary – all ages about the same, wealthier countries.

4. What was the sex ratio for India in 1991 and 2001? What factors shape the sex ratio in India? Why focus on the sex ratio for the 0-6 age group?

5. Why is India's mortality rate lower than the rate of the United States?

less old people

6. Distinguish between internal and international migration.

internal – movement of people within a country (one state to another)

international – movement between countries

7. At what point in history did the world's population reach 1 billion? How long did it take to reach 2 billion? 3 billion? 6 billion?

8. When referring to countries, how is the dichotomy "industrialized—industrializing" misleading?

implies that a failure to industrialize is what makes that country poor.

9. How are labor intensive economies different from core economies?

"Third world" - have higher doubling time, more infant deaths, less GDP income, consume less electricity.

10. Why is Stage 1 of the demographic transition called the stage of "high potential growth"?

high birth, high death

11. According to the model of the demographic transition, which factors contributed to a decline in the death rate? To a rise and than an eventual decline in fertility?

medicine

birth control.

12. Why does the Demographic Transition model not apply to India and other labor intensive countries?

13. What factors contribute to declines in total fertility? To what extent has India realized these factors?

14. What is a demographic trap?

The point at which pop growth overwhelms the environments carrying capacity.

15. What is urbanization?

increase in the pop living in cities.

16. What is a mega city?

8million people + urban agglomeration

17. How does urbanization in labor-intensive poor countries differ from urbanization in core economies?

18. Distinguish among a central city, a suburb, and a non metropolitan area.

Metropolitan area – 1+ cities with 50,000+ residents surrounded by counties

Central city – largest city in metro sometimes 2+

Suburb – concentrated district just outside city boundaries

non-metro – beyond political boundaries.

Key Concepts

Demography	pg. 448
Birth rate (crude)	pg. 448
Age-specific birth rate	pg. 448
Total fertility rate	pg. 449
Death rate (crude)	pg. 449
Infant mortality	pg. 449
Population pyramid	pg. 450
Cohort	pg. 450
Expansive pyramids	pg. 450
Constrictive pyramids	pg. 450
Stationary pyramids	pg. 450
Migration	pg. 453
Migration rate	pg. 453
Push factors	pg. 453
Pull factors	pg. 453
Emigration	pg. 454
Immigration	pg. 455
In-migration	pg. 455
Out-migration	pg. 455
Internal migration	pg. 455
Natural increase	pg. 458
Rate of natural increase	pg. 458
Doubling time	pg. 458
Mortality crisis	pg. 460
Positive checks	pg. 461
Demographic gap	pg. 462
Labor-intensive poor economies	pg. 463
Core economies	pg. 464
Demographic trap	pg. 467
Externality costs	pg. 468
Urbanization	pg. 462
Mega city	pg. 470
Urban agglomeration	pg. 468
Metropolitan statistical area	pg. 471
Central city	pg. 472
Suburb	pg. 175
Nonmetropolitan area	pg. 472

Concept Application

Consider the concepts listed below. Match one or more of the concepts with each scenario. Explain your choices.

- ✓ Cohort
- ✓ Demographic trap
- ✓ Internal migration
- ✓ Migration
- ✓ Positive checks
- ✓ Pull factors
- ✓ Push factors
- ✓ Stationary pyramids
- ✓ Urbanization

Scenario 1 "By 2025, over 1 billion people in Africa and southern Asia will live under conditions of water scarcity. Many North African and Middle Eastern countries are already faced with absolute water scarcity. In Jordan and Israel, over 3,000 people compete for every flow unit of renewable water. By 2025, virtually all North African countries will be faced with high levels of population pressure on their scarce water resources. And, except for Turkey, all of Western Asia will also experience the highest levels of water scarcity" (Falkenmark and Widstrand 1992:20).

Scenario 2 "The reality is of course that, since World War II, tens of millions of people have opted to leave the quiet of the countryside, either "expelled" by drought, disease, or political strife or drawn by dreams broadcast over transistor radios. Some, like the half-million Guatemalan Indians who travel each winter with their wives and families to the Pacific lowlands to pick coffee and cotton or to cut sugarcane do so in order to survive in their villages during the rest of the year. But for most migration is a one-way experience, because those who break with their families and communities, their traditional language, clothes, and food, change too much to be able to return" (Riding 1986:8).

Scenario 3 The population pyramid for Denmark looks more like a rectangle than a pyramid. "Each cohort is about the same size as every other one because the birth rate and the death rate have been low and relatively constant for a long time. These means that each age group is about the same size at birth and, since relatively few people die before old age, the cohorts remain close in size until late in life when mortality rates must rise and eat away at the top of the rectangle" (McFalls 1991:22-23).

Scenario 4 "The villages were as quiet as death.... In one village I remember we had as our guide a tall, middle-aged peasant who had blue eyes and a straw-colored beard. When he spoke of the famine in all those villages hereabouts he struck his breast and tears came into his eyes. He led us into timbered houses where Russian families were hibernating and waiting for death. In some of them they had no food of any kind. There was one family I saw who left an indelible mark on my mind. The father and mother were lying on the floor when we entered and were almost too weak to rise. Some young children were on a bed above the stove, dying of hunger. A boy of eighteen lay back in a wooden settle against the window sill in a kind of coma. These people had nothing to eat—nothing at all" (Gibbs 1987:494).

Scenario 5 "The theme of this book is the lives and reactions of certain patients in a unique situation—and the implications which these hold out for medicine and science. These patients are among the few survivors of the great sleeping-sickness epidemic fifty years ago, and their reactions are those brought about by a remarkable new 'awakening' drug (L-Dopa). The lives and responses of these patients, which have no real precedent in the entire history of medicine, are presented in the form of extended case histories or biographies" (Sacks 1989:1).

Applied Research

Go to the following Census Bureau webpage (http://www.census.gov). Find the World POPClock which gives second by second increases in the world population. Also check out the link "World Vital Events per Time Unit" which shows the number of births and deaths per year, month, day, hour, minute, and second.

Practice Test: Multiple-Choice Questions

1. India's population reached 1 billion in May, 2000. The country might have reached this milestone in 1989 had it not been
   a. for the devastating natural disasters in the past 20 years.
   b. the first country in the world to adopt a national family planning program.
   c. the first country in the world to close its doors to immigrants.
   d. for a history of emigration.

2. The United States emits ____________ percent of all carbon monoxide, the gas responsible for the greenhouse effect.
   a. 5
   b. 15
   c. 24
   d. 50

3. The United States accounts for _________ percent of the world's population.
   a. 30
   b. 10.5
   c. 4.6
   d. 1.9

4. Use the following information to calculate the age specific birth rate for India. Total births in year: 24,424,724; number of women 15-54: 319,259,867. The age-specific birth rate is _________ per 1,000.
   a. 51.99
   b. 5199
   c. 76.5
   d. 109.5

5. A subspecialty within sociology that focuses on the study of human population is
   a. epidemiology.
   b. ethnomethodology.
   c. demography.
   d. conflict theory.

6. A population's age and sex composition is commonly depicted as a
   a. three-dimensional graph.
   b. cohort.
   c. population pyramid.
   d. demographic transition.

7. Country Y has a population of 149.3 million people. Life expectancy is 73 years for men and 79 years for women. The total fertility is below replacement level. The population pyramid for this country would be
   a. expansive.
   b. constrictive.
   c. stationary.
   d. triangular.

8. Stewart is moving out of his hometown because there are no jobs. The reason he is moving is called a
   a. push factor.
   b. pull factor.
   c. demographic.
   d. self-motivating factor.

9. Within the United States, the greatest amount of internal migration is movement
   a. within the same county.
   b. from one county to another.
   c. from one state to another.
   d. into adjacent countries.

10. Almost 30 percent of India's total population changed residences in the past 10 years. This movement is dominated by short-distance rural-to-rural movements within India. Sociologists classify this kind of migration as
    a. immigration.
    b. emigration.
    c. internal migration.
    d. international.

11. An Inter-American Development Bank poll found that almost ___________ adult Mexican residents receive money from relatives working in the United States.
    a. one in twenty
    b. one in ten
    c. one in five
    d. one in two

12. The demographic transition
    a. is a two-stage model of population growth.
    b. depicts the history of birth and death rates in labor-intensive poor countries.
    c. depicts the history of disease in mechanized-rich countries.
    d. depicts the history of population growth in Western Europe and North America.

13. 50/1000 is believed to be the highest ______ rate possible for any society.
    a. death
    b. fertility
    c. marriage
    d. birth

14. In demographic terms, the Black Death is an example of
    a. a mortality crisis.
    b. a life expectancy crisis.
    c. a tragedy.
    d. a degenerative disease.

15. According to Thomas Malthus, epidemics, war, and famine are examples of
    a. positive checks.
    b. demographic traps.
    c. demographic gaps.
    d. catastrophic events.

16. The least important reason for the decline in death rates in Western societies is
    a. improvements in agricultural technology.
    b. improvement in sanitation.
    c. medical advances.
    d. proper disposal of sewage.

17. Which one of the following countries is least likely to be in stage 3 of the Demographic Transition?
    a. United States
    b. Germany
    c. India
    d. Japan

18. Urbanization includes all but which one of the following characteristics?
    a. increase in the number of cities
    b. growth of the population living in cities
    c. rural-to-urban migration
    d. urban-to-rural migration

19. The __________ is considered "the most illustrious and most flourishing commercial association that ever existed in any age or country."
    a. Bank of India
    b. General Electric Company
    c. East India Company
    d. Domino Sugar Company

20. India was once a colony of
    a. the United States.
    b. Portugal.
    c. Spain.
    d. Britain.

21. Which one of the following cities in the country of India is not considered a mega city?
    a. Bombay
    b. Calcutta
    c. Delhi
    d. Bhopal

22. Which one of the following U.S. cities is not among the world's top 30 economies?
    a. San Francisco
    b. Boston
    c. Houston
    d. Philadelphia

23. The most notable characteristic of housing units in nonmetropolitan areas is that a significant percentage are
    a. more than 50 years old.
    b. within 300 feet of a commercial establishment.
    c. renter-occupied.
    d. mobile homes.

24. One important trend helping in India's effort to promote lower fertility is that
    a. almost 70 percent of India's labor force is employed in agriculture.
    b. life expectancy is 55 years of age.
    c. almost all women in India are married by 19.
    d. 60 percent of women are enrolled in school.

True/False Questions

T F 1. China is the country with the largest population in the world.

T F 2. Japan is among the 10 most populous countries in the world.

T F 3. Historically, emigration rates for India were especially low during times of major famine and epidemics.

T F 4. In at least one state in India the sex ratio is 783 females per 1,000 males.

T F 5. Humans produce enough food each year to feed the world's population.

T F 6. The Industrial Revolution was an event confined to the world's core economies.

T F 7. In India, female sterilization appears to be a major form of contraception.

T F 8. In India, total fertility has increased since 1960.

Internet Resources Related to Population

- **Population Reference Bureau**
  http://www.prb.org
  The Population Reference Bureau covers births, deaths, migration, and other population topics as they relate to the United States and other countries. Examples of the hundreds of articles on the webpage include "Why Do Canadians Outlive Americans?," "The Lives and Times of the Baby Boomers," and "Hurricanes, Population Trends, and Environmental Change."

- **United Nations Populations Fund**
  http://www.unfpa.org/index.htm
  "The United Nations Population Fund extends assistance to developing countries, countries with economies in transition and other countries at their request to help them address reproductive health and population issues, and raises awareness of these issues in all countries." This site contains information on improving reproductive health, preventing HIV infection and other population issues for all countries.

Internet Resources Related to India

- **The *Times of India***
  http://timesofindia.indiatimes.com
  The *Times of India* covers major news and current events relevant to India and India's relationship to the world.

- **Census of India**
  http://www.censusindia.net/
  Census data, maps and vital statistics about India can be found on this site from the Office of the Registrar General, India.

India: Background Notes

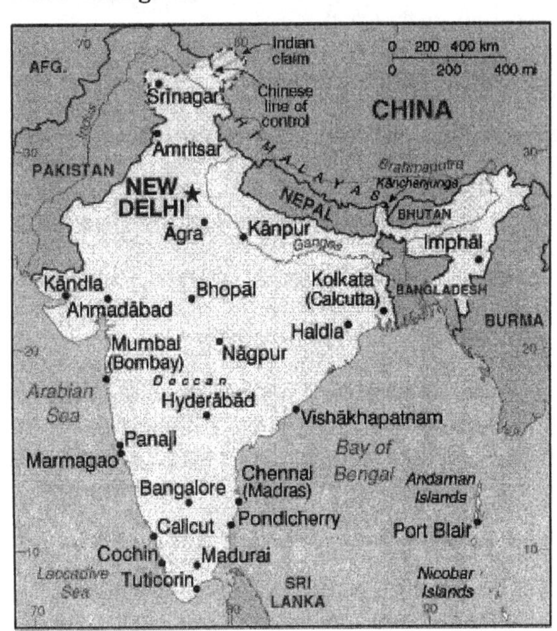

PEOPLE

Nationality: *Noun and adjective*--Indian(s).
Population (2003 est.): 1.05 billion; urban 27.8%.
Annual growth rate: 1.6%.
Density: 319/sq. km.
Ethnic groups: Indo-Aryan 72%, Dravidian 25%, Mongoloid 2%, others.
Religions: Hindu 81.3%, Muslim 12%, Christian 2.3%, Sikh 1.9%, other groups including Buddhist, Jain, Parsi 2.5%.
Languages: Hindi, English, and 16 other official languages.
Education: *Years compulsory*--9 (to age 14). *Literacy*--55.2%.
Health: *Infant mortality rate*--61/1,000. *Life expectancy*--63 years.
Work force (est.): 416 million. *Agriculture--63%; industry and commerce*--22%; *services and government*--11%; *transport and communications*--4%.

Although India occupies only 2.4% of the world's land area, it supports over 15% of the world's population. Only China has a larger population. Almost 33% of Indians are younger than 15 years of age. About 70% of the people live in more than 550,000 villages, and the remainder in more than 200 towns and cities. Over thousands of years of its history, India has been invaded from the Iranian plateau, Central Asia, Arabia, Afghanistan, and the West; Indian people and culture have absorbed and changed these influences to produce a remarkable racial and cultural synthesis.

Religion, caste, and language are major determinants of social and political organization in India today. The government has recognized 18 languages as official; Hindi is the most widely spoken.
Although 81% of the people are Hindu, India also is the home of more than 126 million Muslims--one of the world's largest Muslim populations. The population also includes Christians, Sikhs, Jains, Buddhists, and Parsis.

The caste system reflects Indian occupational and socially defined hierarchies. Sanskrit sources refer to four social categories, priests (Brahmin), warriors (kshatriya), traders (vayisha) and farmers (shudra). Although these categories are understood throughout India, they describe reality only in the most general terms. They omit, for example, the tribes and low castes once known as 'untouchables.' In reality, society in India is divided into thousands of jatis, local, endogamous groups, organized hierarchically according to complex ideas of purity and pollution. Despite economic modernization and laws countering discrimination against the lower end of the class structure, the caste system remains an important source of social identification for most Hindus and a potent factor in the political life of the country.

HISTORY

The people of India have had a continuous civilization since 2500 B.C., when the inhabitants of the Indus River valley developed an urban culture based on commerce and sustained by agricultural trade. This civilization declined around 1500 B.C., probably due to ecological changes.

During the second millennium B.C., pastoral, Aryan-speaking tribes migrated from the northwest into the subcontinent. As they settled in the middle Ganges River valley, they adapted to antecedent cultures.

The political map of ancient and medieval India was made up of myriad kingdoms with fluctuating boundaries. In the 4th and 5th centuries A.D., northern India was unified under the Gupta Dynasty. During this period, known as India's Golden Age, Hindu culture and political administration reached new heights.

Islam spread across the subcontinent over a period of 500 years. In the 10th and 11th centuries, Turks and Afghans invaded India and established sultanates in Delhi. In the early 16th century, the Chaghtai Turkish adventurer and distant relative of Timurlang, Babur, established the Mughal Dynasty, which lasted for 200 years. South India followed an independent path, but by the 17th century it too came under the direct rule of influence of the expanding Mughal Empire. While most of Indian society in its thousands of villages remained untouched by the political struggles going on around them, Indian courtly culture evolved into a unique blend of Hindu and Muslim traditions.

The first British outpost in South Asia was established by the English East India Company in 1619 at Surat on the northwestern coast. Later in the century, the Company opened permanent trading stations at Madras, Bombay, and Calcutta, each under the protection of native rulers.

The British expanded their influence from these footholds until, by the 1850s, they controlled most of present-day India, Pakistan, and Bangladesh. In 1857, a rebellion in north India led by mutinous Indian soldiers caused the British Parliament to transfer all political power from the East India Company to the Crown. Great Britain began administering most of India directly while controlling the rest through treaties with local rulers.

In the late 1800s, the first steps were taken toward self-government in British India with the appointment of Indian councilors to advise the British viceroy and the establishment of provincial councils with Indian members; the British subsequently widened participation in legislative councils. Beginning in 1920, Indian leader Mohandas K. Gandhi transformed the Indian National Congress political party into a mass movement to campaign against British colonial rule. The party used both parliamentary and nonviolent resistance and noncooperation to achieve independence.

On August 15, 1947, India became a dominion within the Commonwealth, with Jawaharlal Nehru as Prime Minister. Enmity between Hindus and Muslims led the British to partition British India, creating East and West Pakistan, where there were Muslim majorities. India became a republic within the Commonwealth after promulgating its Constitution on January 26, 1950.

After independence, the Congress Party, the party of Mahatma Gandhi and Jawaharlal Nehru, ruled India under the influence first of Nehru and then his daughter and grandson, with the exception of two brief periods in the 1970s and 1980s.

Prime Minister Nehru governed the nation until his death in 1964. He was succeeded by Lal Bahadur Shastri, who also died in office. In 1966, power passed to Nehru's daughter, Indira Gandhi, Prime Minister from 1966 to 1977. In 1975, beset with deepening political and economic problems, Mrs. Gandhi declared a state of emergency and suspended many civil liberties. Seeking a mandate at the polls for her policies, she called for elections in 1977, only to be defeated by Moraji Desai, who headed the Janata Party, an amalgam of five opposition parties.

In 1979, Desai's Government crumbled. Charan Singh formed an interim government, which was followed by Mrs. Gandhi's return to power in January 1980. On October 31, 1984, Mrs. Gandhi was assassinated, and her son, Rajiv, was chosen by the Congress (I)--for "Indira"--Party to take her place. His Congress government was plagued with allegations of corruption resulting in an early call for national elections in 1989.

In the 1989 elections Rajiv Gandhi and Congress won more seats than any other single party, but he was unable to form a government with a clear majority. The Janata Dal, a union of opposition parties, then joined with the Hindu-nationalist Bharatiya Janata Party (BJP) on the right and the communists on the left to form the government. This loose coalition collapsed in November 1990, and Janata Dal, supported by the Congress (I), came to power for a short period, with Chandra Shekhar as Prime Minister. That alliance also collapsed, resulting in national elections in June 1991.

On May 27, 1991, while campaigning in Tamil Nadu on behalf of Congress (I), Rajiv Gandhi was assassinated, apparently by Tamil extremists from Sri Lanka. In the elections, Congress (I) won 213 parliamentary seats and returned to power at the head of a coalition, under the leadership of P.V. Narasimha Rao. This Congress-led government, which served a full 5-year term, initiated a gradual process of economic liberalization and reform, which opened the Indian economy to global trade and investment. India's domestic politics also took new shape, as the nationalist appeal of the Congress Party gave way to traditional alignments by caste, creed, and ethnicity leading to the founding of a plethora of small, regionally based political parties.

The final months of the Rao-led government in the spring of 1996 were marred by several major political corruption scandals, which contributed to the worst electoral performance by the Congress Party in its history. The Hindu-nationalist Bharatiya Janata Party (BJP) emerged from the May 1996 national elections as the single-largest party in the Lok Sabha but without a parliamentary majority. Under Prime Minister Atal Bihari Vajpayee, the subsequent BJP coalition lasted only 13 days. With all political parties wishing to avoid another round of elections, a 14-party coalition led by the Janata Dal formed a government known as the United Front, under the former Chief Minister of Karnataka, H.D. Deve Gowda. His government collapsed after less than a year, when the Congress Party withdrew his support in March 1997. Inder Kumar Gujral replaced Deve Gowda as the consensus choice for Prime Minister at the head of a 16-party United Front coalition.

In November 1997, the Congress Party again withdrew support from the United Front. In new elections in February 1998, the BJP won the largest number of seats in Parliament--182--but fell far short of a majority. On March 20, 1998, the President inaugurated a BJP-led coalition government with Vajpayee again serving as Prime Minister. On May 11 and 13, 1998, this government conducted a series of underground nuclear tests, forcing U.S. President Clinton to impose economic sanctions on India pursuant to the 1994 Nuclear Proliferation Prevention Act.

In April 1999, the BJP-led coalition government fell apart, leading to fresh elections in September. The National Democratic Alliance--a new coalition led by the BJP--gained a majority to form the government with Vajpayee as Prime Minister in October 1999.

The Kargil conflict in 1999 and an attack on the Indian Parliament in December 2001 led to increased tensions with Pakistan. Hindu nationalists have long agitated to build a temple on a disputed site in Ayodhya. In February 2002, a mob of Muslims attacked a train carrying Hindu volunteers returning from Ayodhya to the state of Gujarat, and 57 were burnt alive. Over 900 people were killed and 100,000 left homeless in the resulting anti-Muslim riots throughout the state. This led to accusations that the state government had not done enough to contain the riots, or arrest and prosecute the rioters.

The ruling BJP-led coalition was defeated in a five-stage election held in April and May of 2004, and a Congress-led coalition took power on May 22 with Manmohan Singh as Prime Minister.

ECONOMY

GDP: $576 billion (2003); $648 billion (2004 est.).
Real growth rate: 8.2% (2003).
Per capita GDP: $543 (2003); $602 (2004 est.).
Natural resources: Coal, iron ore, manganese, mica, bauxite, chromite, thorium, limestone, barite, titanium ore, diamonds, crude oil.
Agriculture: 22.7% of GDP. *Products*--wheat, rice, coarse grains, oilseeds, sugar, cotton, jute, tea
Industry: 26.6% of GDP. *Products*--textiles, jute, processed food, steel, machinery, transport equipment, cement, aluminum, fertilizers, mining, petroleum, chemicals, computer software.
Services and transportation: 50.7% of GDP.
Trade: *Exports*--$62 billion; agricultural products, engineering goods, precious stones, cotton apparel and fabrics, gems and jewelry, handicrafts, tea. *Software exports*--$12.5 billion. *Imports*--$76 billion; petroleum, machinery and transport equipment, electronic goods, edible oils, fertilizers, chemicals, gold, textiles, iron and steel. *Major trade partners*--U.S., EU, Russia, Japan, Iraq.

India's population is estimated at nearly 1.07 billion and is growing at 1.7% a year. It has the world's 12th largest economy--and the third largest in Asia behind Japan and China--with total GDP of around $570 billion. Services, industry and agriculture account for 50.7%, 26.6% and 22.7% of GDP respectively. Nearly two-thirds of the population depends on agriculture for their livelihood. About 25% of the population lives below the poverty line, but a large and growing middle class of 320-340 million has disposable income for consumer goods.

India is continuing to move forward with market-oriented economic reforms that began in 1991. Recent reforms include liberalized foreign investment and exchange regimes, industrial decontrol, significant reductions in tariffs and other trade barriers, reform and modernization of the financial sector, significant adjustments in government monetary and fiscal policies and safeguarding intellectual property rights.

Real GDP growth for the fiscal year ending March 31, 2004 was 8.17%, up from the drought-depressed 4.0% growth in the previous year. Growth for the year ending March 31, 2005 is expected to be between 6.5% and 7.0%. Foreign portfolio and direct investment in-flows have risen significantly in recent years. They have contributed to the $120 billion in foreign exchange reserves at the end of June 2004. Government receipts from privatization were about $3 billion in fiscal year 2003-04.

However, economic growth is constrained by inadequate infrastructure, a cumbersome bureaucracy, corruption, labor market rigidities, regulatory and foreign investment controls, the "reservation" of key products for small-scale industries and high fiscal deficits. The outlook for further trade liberalization is mixed. India eliminated quotas on 1,420 consumer imports in 2002 and has announced its intention to continue to lower customs duties. However, the tax structure is complex with compounding effects of various taxes.

The United States is India's largest trading partner. Bilateral trade in 2003 was $18.1 billion and is expected to reach $20 billion in 2004. Principal U.S. exports are diagnostic or lab reagents, aircraft and parts, advanced machinery, cotton, fertilizers, ferrous waste/scrap metal and computer hardware. Major U.S. imports from India include textiles and ready-made garments, internet-enabled services, agricultural and related products, gems and jewelry, leather products and chemicals.

The rapidly growing software sector is boosting service exports and modernizing India's economy. Revenues from IT industry are expected to cross $20 billion in 2004-05. Software exports were $12.5 billion in 2003-04. PC penetration is 8 per 1,000 persons, but is expected to grow to 10 per 1,000 by 2005. The cellular mobile market is expected to surge to over 50 million subscribers by 2005 from the present 36 million users. The country has 52 million cable TV customers.

The United States is India's largest investment partner, with total inflow of U.S. direct investment estimated at $3.7 billion in 2003. Proposals for direct foreign investment are considered by the Foreign Investment Promotion Board and generally receive government approval. Automatic approvals are available for investments involving up to 100% foreign equity, depending on the kind of industry. Foreign investment is particularly sought after in power generation, telecommunications, ports, roads, petroleum exploration/processing and mining.

India's external debt was $112 billion in 2003, up from $105 billion in 2002. Bilateral assistance was approximately $2.62 billion in 2002-03, with the United States providing about $130.2 million in development assistance in 2003. The World Bank plans to double aid to India to almost $3 billion over the next four years, beginning in July 2004.

U.S.-INDIA RELATIONS

The United States has undertaken a transformation in its relationship with India based on the conviction that U.S. interests require a strong relationship with India. The two countries are the largest democracies, committed to political freedom protected by representative government. India is also moving toward greater economic freedom. The two have a common interest in the free flow of commerce, including through the vital seas lanes of the Indian Ocean. They also share an interest in fighting terrorism and in creating a strategically stable Asia.

Differences remain, including over India's nuclear weapons programs and over the pace of India's economic reforms. But while in the past these concerns may have dominated U.S. thinking about India, today the U.S. starts with a view of India as a growing world power with which it shares common strategic interests. Through a strong partnership with India, the two countries can best address differences and shape a dynamic future.

In late September 2001, President Bush lifted the sanctions that were imposed under the terms of the 1994 Nuclear Proliferation Prevention Act following India's nuclear tests in May 1998. The nonproliferation dialogue initiated after the 1998 nuclear tests has bridged many of the gaps in understanding between the countries. President Bush met Prime Minister Vajpayee in November 2001, and the two leaders expressed a strong interest in transforming the U.S.-India bilateral relationship. High-level meetings and concrete cooperation between the two countries increased during 2002 and 2003. The U.S. and India announced on January 12, 2004, the Next Steps in Strategic Partnership (NSSP), both a milestone in the transformation of the bilateral relationship and a blueprint for its further progress. Progress has been made on this initiative with the conclusion of Phase I in late September 2004.

Source: U.S. Department of State (2004)
http://www.state.gov

Chapter References

Brenson, Michael. 1991. "Images of People Who Live Outside Power." *The New York Times* (April 5):B1.

Falkenmark, Malin and Carl Widstrand. 1992. "Population and Water Resources: a Delicate Balance." *Population Bulletin* 47(3):1-35.

Gibb, Philip. 1987. "Famine in Russia, October 1921." Pp. 493-95 in *Eyewitness to History*, edited by J. Carey. Cambridge: Harvard University Press.

Haub, Carl and Machiko Yanagishita. 1994. *1994 World Population Data Sheet*. Washington, DC: Population Reference Bureau.

McFalls, Joseph A., Jr. 1991. "Population: A Lively Introduction." Washington, DC: *Population Bulletin* 46(2): 1-40..

Riding, Alan. 1986. "Introduction." Pp. 7-9 in *Other Americas*, by Sebastiao Salgado. New York: Pantheon.

Sacks, Oliver. 1983. *Awakenings*. New York: Dutton.

U.S. Department of State. 1990. "Brazil." *Background Notes* (#7756). Washington, DC: U.S. Government Printing Office.

Answers

Concept Application

1. Demographic trap	pg. 467
2. Urbanization; Migration; Push factors; Pull factors; Internal migration	pg. 462; 453; 453; 453
3. Stationary pyramid	pg. 450
4. Positive checks	pg. 461
5. Cohort	pg. 450

Multiple-Choice

1. b	pg. 446	13. d	pg. 461
2. c	pg. 447	14. a	pg. 460
3. c	pg. 447	15 a	pg. 461
4. c	pg. 448	16. c	pg. 462
5. c	pg. 448	17. c	pg. 462
6. c	pg. 450	18. d	pg. 462
7. b	pg. 450	19. c	pg. 464
8. a	pg. 453	20. d	pg. 464
9. a	pg. 454	21. d	pg. 470
10. c	pg. 454	22. a	pg. 473
11. c	pg. 455	23. d	pg. 473
12. d	pg. 459	24. d	pg. 474

True/False

1. T	pg. 447
2. T	pg. 448
3. F	pg. 454
4. T	pg. 453
5. T	pg. 467
6. F	pg. 464
7. T	pg. 465
8. F	pg. 474

Chapter 14

Education
With Emphasis on the European Union

Study Questions

1. Why was the education system of the European Union chosen as the emphasis for the Education chapter?

2. What impressions of early American education did Europeans have?

3. Distinguish between schooling and formal education.

4. What are the two major conceptions of the functions of schooling?

5. Expand on the statement "Illiteracy is a product of one's environment."

6. What factors encouraged parents to comply with compulsory attendance laws when they were first instituted in the United States?

7. What features of early American education survive today? Explain how they are revealed in homework and class assignments.

8. List the fundamental characteristics of American education. Briefly explain how each is related to problems in American education.

9. How does the United States compare with the European Union on providing its population opportunities to attend college?

10. Distinguish between formal and hidden curriculum.

11. What is "spelling baseball"? What do children learn when they engage in such educational activities? Why?

12. What is tracking? What is the rationale for tracking? Is this rationale supported by research? Why does tracking persist?

13. Explain how the self-fulfilling prophecy can affect students' academic achievements.

14. What did James Coleman uncover about American schools? What was the most controversial finding?

15. How were Coleman's findings "used"? What happened when his recommendation to bus students was implemented?

16. According to Coleman, how did the adolescent subculture emerge?

17. What are the major characteristics of the adolescent status system? How does it reflect values of the society? How does it affect education?

18. What evidence exists to support the European assessment that U.S. students are preoccupied with knowledge as it applies to income generation and wealth creation?

19. What evidence exists to support the European observation that U.S. students value personal observations over accumulated knowledge and direct experience with other ways of life?

20. Is there evidence to support the European observation that the ideal person is self-made?

21. Is there evidence to support the European claim that U.S. students place high value on educational achievement but not on the dedicated study needed to achieve it?

Key Concepts

Education	
Informal education	pg. 481
Formal education	pg. 481
Schooling	pg. 481
Functionally illiterate	pg. 482
Illiteracy	pg. 482
Formal curriculum	pg. 494
Hidden curriculum	pg. 495
Self-fulfilling prophecy	pg. 501

Concept Application

Consider the concepts listed below. Match one or more of the concepts with each scenario. Explain your choices.

- ✓ Ability grouping/Tracking
- ✓ Formal education
- ✓ Functionally illiterate
- ✓ Informal education
- ✓ Schooling
- ✓ Self-fulfilling prophecy
- ✓ Social promotion

Scenario 1 "Many of the deaf are functional illiterates.... Hans Furth, a psychologist whose work is concerned with the cognition of the deaf,... argues that the congenitally deaf suffer from 'information deprivation.' There are a number of reasons for this. First, they are less exposed to the 'incidental' learning that takes place out of school—for example, to that buzz of conversation that is the background of ordinary life; to television, unless it is captioned, etc. Second, the content of deaf education is meager compared to that of hearing children; so much time is spent teaching deaf children speech—one must envisage between five and eight years of intensive tutoring—that there is little time for transmitting information, culture, complex skills, or anything else.

Yet the desire to have the deaf speak, the insistence that they speak—and from the first, the odd superstitions that have always clustered around the use of sign language, to say nothing of the enormous investment in oral schools allowed this deplorable situation to develop, practically unnoticed except by deaf people, who themselves being unnoticed had little to say in the matter" (Sacks 1989:28-29).

Scenario 2 "In 1897, Captain Richard Pratt arrived in Sioux country to enlist Sioux children for his Carlisle Indian Industrial School, the first and most famous of what would become a whole system of off-reservation boarding schools for Indian students. Eighty-four Sioux children from Pine Ridge and Rosebud, about two-thirds boys and mainly from prominent families, returned east with the stern captain. Neither parent nor pupil foresaw the short hair, the starched shirts and squeaky boots, the Christian names, or the other trappings.... Head shaving and even shackling with a ball and chain were common punishments for Indian pupils who ran away or spoke in their native tongue. Suppressing the Sioux language was rated high among both the Indian Bureau's educational priorities and the reasons Sioux parents kept children at home" (Lazarus 1991:101-03).

Scenario 3 "Given a paycheck and the stub that lists the usual deductions, 26 percent of adult Americans cannot determine if their paycheck is correct. Thirty-six percent, given a W-4 form, cannot enter the right numbers of exemptions in the proper places on the form. Forty-four percent, when given a series of 'help-wanted' ads, cannot match their qualifications to the job requirements. Twenty-two percent cannot address a letter well enough to guarantee that it will reach its destination. Twenty-four percent cannot add their own correct return address to the same envelope. Twenty percent cannot understand an 'equal opportunity' announcement. Over 60 percent, given a series of 'for sale' advertisements for products new and used, cannot calculate the difference between prices for a new and used appliance" (Kozol 1985:9).

Scenario 4 "The development of IQ tests lent an air of objectivity to the placement of procedures used to separate children for instruction....Test pioneer Lewis Terman wrote in 1916: 'At every step in the child's progress the school should take account of his vocational possibilities. Preliminary investigations indicate that an IQ below 70 rarely permits anything better than unskilled labor; that range from 70 to 80 is pre-eminently that of semi-skilled labor, from 80 to 100 that of skilled or ordinary clerical labor, from 100 to 110 or 115 that of the semi-professional pursuits; and that above these are the grades of intelligence which permit one to enter the professions or the larger fields of business....This information will be a great value in planning the education of a particular child and also in planning the differentiated curriculum here recommended.'" (Oakes 1985, p.36).

Applied Research

Go to the Council of the Great City Schools website to find the report What Works in Urban Education. The report shares descriptions of 155 successful programs. List 4-5 strategies that work to enhance the educational experience.

Practice Test: Multiple-Choice Questions

1. In the broadest sense, education is
   a. purposeful, planned effort to impart specific skills.
   b. a program of formal and systematic instruction.
   c. those experiences that train, discipline, and develop mental and physical potentials.
   d. spontaneous, unplanned exposure to ideas.

2. Early antidotal evidence from British observers suggests that the U.S. system of education seems to value
   a. dedicated study.
   b. income generation and wealth creation.
   c. experience with other ways of life.
   d. accumulated knowledge.

3. We focus on the European Union in the Education chapter for all but which one of the following reasons?
   a. The EU is investing heavily in education and research to boost its international competitiveness.
   b. The EU is emphasizing higher education's role in preparing its citizens to live and work in different European cultures.
   c. The EU is working to limit opportunities for education beyond high school.
   d. EU openness to attracting international students is expected to challenge U.S. dominance as a host country.

4. Which one of the following characteristics applies to the process of informal education?
   a. purposeful
   b. systematic
   c. spontaneous
   d. planned

5. Sociologist Emile Durkheim suggests that schools function
   a. to meet the needs of society.
   b. to create minds critical of current social arrangements.
   c. as marriage markets.
   d. to liberate people from the past.

6. The contextual nature of illiteracy suggests that it is
   a. like a disease.
   b. linked to a lack of desire to want to read and write.
   c. biologically rooted.
   d. a social phenomenon.

7. Students in 7 of 19 EU countries score higher than U.S. students in science literacy tests. Which one of the following countries is among these seven?
   a. Italy
   b. Spain
   c. Finland
   d. Poland

8. Approximately _______ percent of ninth graders enrolled in U.S. public schools graduate from high school four years later.
   a. 25
   b. 70
   c. 90
   d. 95

9. Two of the most prominent features of early American education that have endured to the present are
   a. tracking and dull assignments.
   b. spelling baseball and dodge ball.
   c. textbooks modeled after catechisms and single-language instruction.
   d. mass education and integration.

10. In the United States 63 percent of high school graduates enroll in college the following year. The figure would decrease to ______ percent if we considered the percentage of all 18 year olds.
    a. 20
    b. 40
    c. 50
    d. 60

11. In ________________ models school districts and/or local authorities establish the curriculum.
    a. minimalist
    b. national
    c. decentralized
    d. centralized

12. Although all countries in the world have education-based programs that address social problems, __________ is unique in that education is viewed as the primary solution to many of its problems.
    a. the United States
    b. Japan
    c. Canada
    d. Mexico

13. Teaching method, tone of teacher's voice, and frequency of teacher's absences fall under the category of __________ curriculum.
    a. formal
    b. hidden
    c. unintended
    d. planned

14. Jules Henry argues that students go along with teachers' requests and participate in activities such as "spelling baseball" because they
    a. don't care whether they learn or not.
    b. are terrified of failure and want so badly to succeed.
    c. find such academic "games" entertaining.
    d. find the competition enjoyable.

15. Sociologist Jeannie Oakes studied a wide range of school systems and came to the conclusion that __________ has/have the greatest effect on quality of education.
    a. tracking
    b. rural-urban environments
    c. type of school (public versus private)
    d. amount of cultural diversity

16. Which one of the following responses to the question "What is the most important thing you learned or have done so far in class?" is the one mostly likely to be made by a student placed in college preparatory track?
    a. "I think the most important thing I learned so far is to come into math class and get out folders."
    b. "I learned to be more imaginative."
    c. "I have learned nothing that I'd use in later life."
    d. "To be honest, I have learned nothing."

17. ________________ programs prepare students for direct entry into a specific occupation.
    a. College preparatory
    b. Transitional
    c. Vocational
    d. General studies

18. In comparison to their American counterparts, Asian teachers
    a. work together very closely in preparing lesson plans.
    b. are in the classroom for a smaller portion of the day.
    c. work in environments that discourage systematic learning by students outside the classroom.
    d. receive more systematic training, which prepares them to deal with an array of discipline problems.

19. Which of the following is <u>not</u> one of the major findings reported in "Equality of Educational Opportunity"?
    a. Variations in the quality of a school did not have much effect on student test scores.
    b. The social class of one's classmates had a significant effect on student test scores.
    c. School expenditures are an important predictor of educational attainment.
    d. Schools bring little to bear on a child's achievement independent of the child's immediate environment.

20. The Coleman report and other studies have found that ______ is the most powerful factor in determining the level of school achievement of students.
    a. race
    b. social class
    c. home environment (e.g., family background)
    d. gender

21. Coleman's study of adolescent subcultures has this important implication for understanding why even the best students in the United States have difficulty in competing with top students in many other countries:
    a. The United States does not draw into competition everyone who has academic potential.
    b. The United States gives everyone access to education.
    c. Parents no longer "train" their children in the skills they know because those skills are outdated and obsolete.
    d. Parents exercise more influence than teachers over their children's lives.

22. Sociologist James Coleman defined the ________ as a "small society, one that has most of its important interactions within itself, and maintains only a few threads of connection with the outside adult society."
    a. white flight
    b. schooling
    c. adolescent subculture
    d. formal education

23. Students who perceive borders as insurmountable and immerse themselves in the world of their peers would be classified as
    a. congruent worlds/smooth transitions.
    b. different worlds/border crossings managed.
    c. different worlds/border crossings difficult.
    d. different worlds/border crossings resisted.

24. _____________ of U.S. college students borrow money to pay for college.
    a. 90
    b. 75
    c. 65
    d. 25

True/False Questions

T F 1. In some European Union countries, mandatory foreign language study begins as early as age six.

T F 2. Japan was the first country in the world to embrace the concept of mass education.

T F 3. The first textbooks in the United States were modeled after catechisms.

T F 4. The relatively unrestricted right to a college education seems to be connected to a decline in the value of a high school diploma.

T F 5. A national or centralized curriculum sets achievement targets but allows schools to set their own curriculum.

T F 6. Within the European Union, the spending gap between the wealthiest and poorest countries is greater than the spending gap between the wealthiest and poorest states.

T F 7. Jules Henry uses the example of math races to illustrate how the hidden curriculum works.

T F 8. Self-fulfilling prophecies begin with a false definition of a situation.

T F 9. The single most important factor in explaining academic success is family background.

T F 10. Relative to other systems of education, the U.S. system seems to promote a math curriculum that makes learning less rewarding and interesting.

Internet Resources Related to Education in the United States

- **Condition of Education**
  http://nces.ed.gov/programs/coe/
  The Condition of Education is an annual report prepared by the U.S. Department of Education National Center for Educational Statistics. The full text of the report is online, and it covers "trends in enrollments, student achievement, dropout rates, degree attainment, long-term outcomes of education, and education financing".

Internet Resources Related to the European Union

- **European Union News**
  http:// eunews.proeurope.org/eu_news.php
  For the latest news on the European Union in general and on EU Institutions and Agencies visit this website.

- **European Union in the US**
  http://www.eurunion.org/
  This European Union website contains information on EU countries, law and policy overviews, research tools and a link about the EU for kids.

The European Union: Background Notes

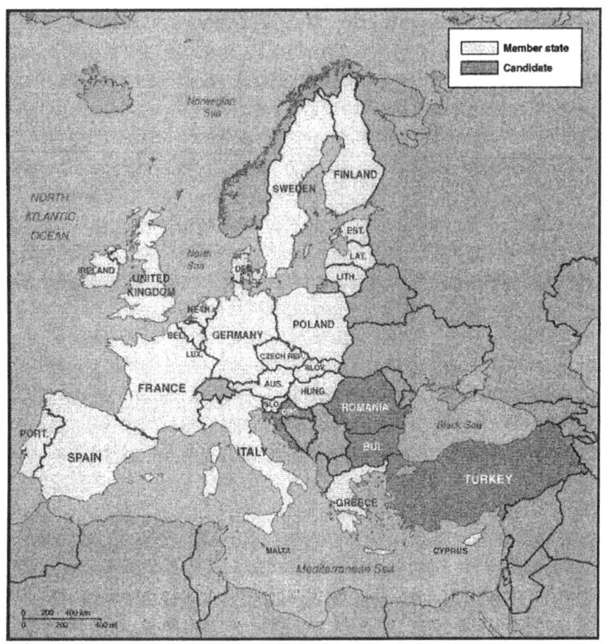

BACKGROUND

Following the two devastating World Wars of the first half of the 20th century, a number of European leaders in the late 1940s became convinced that the only way to establish a lasting peace was to unite the two chief belligerent nations - France and Germany - both economically and politically. In 1950, the French Foreign Minister Robert Schuman proposed an eventual union of all of Europe, the first step of which would be the integration of the coal and steel industries of Western Europe. The following year the European Coal and Steel Community (ECSC) was set up when six members, Belgium, France, West Germany, Italy, Luxembourg, and the Netherlands, signed the Treaty of Paris.

The ECSC was so successful that within a few years the decision was made to integrate other parts of the countries' economies. In 1957, the Treaties of Rome created the European Economic Community (EEC) and the European Atomic Energy Community (EURATOM), and the six member states undertook to eliminate trade barriers among themselves by forming a common market. In 1967, the institutions of all three communities were formally merged into the European Community (EC), creating a single Commission, a single Council of Ministers, and the European Parliament. Members of the European Parliament were initially selected by national parliaments, but in 1979 the first direct elections were undertaken and they have been held every five years since.

In 1973, the first enlargement of the EC took place with the addition of Denmark, Ireland, and the United Kingdom. The 1980s saw further membership expansion with Greece joining in 1981 and Spain and Portugal in 1986. The 1992 Treaty of Maastricht laid the basis for further forms of cooperation in foreign and defense policy, in judicial and internal affairs, and in the creation of an economic and monetary union - including a common currency. This further integration created the European Union (EU). In 1995, Austria, Finland, and Sweden joined the EU, raising the membership total to 15.

A new currency, the euro, was launched in world money markets on 1 January 1999; it became the unit of exchange for all of the EU states except Great Britain, Sweden, and Denmark. In 2002, citizens of the 12 euro-area countries began using euro banknotes and coins. Ten new countries joined the EU in 2004—Cyprus, the Czech Republic, Estonia, Hungary, Latvia, Lithuania, Malta, Poland, Slovakia, and Slovenia—bringing the current membership to 25. In order to ensure that the EU can continue to function efficiently with an expanded membership, the 2003 Treaty of Nice set forth rules streamlining the size and procedures of EU institutions. An EU Constitutional Treaty, signed in Rome on 29 October 2004, gives member states two years to ratify the document before it is scheduled to take effect on 1 November 2006.

Despite the expansion of membership and functions, "Eurosceptics" in various countries have raised questions about the erosion of national cultures and the imposition of a flood of regulations from the EU capital in Brussels. Failure by member states to ratify the constitution or the inability of newcomer countries to meet euro currency standards might force a loosening of some EU agreements and perhaps lead to several levels of EU participation. These "tiers" might eventually range from an "inner" core of politically integrated countries to a looser "outer" economic association of members.

AREA

Total area: 3,976,372 sq km
Land area: 11,214.8 km
Comparative area: Less than one-half the size of the US
Natural resources: Iron ore, arable land, natural gas, petroleum, coal, copper, lead, zinc, hydropower, uranium, potash, fish

PEOPLE

Population: 456,285,839 (July 2004 est.)

Age structure:
0-14 years: 16.3%
15-64 years: 67.2%
65 years and over: 16.6% (July 2004 est.)

Population growth rate: 0.17% (July 2004 est.)

Birth rate: 10.2 births/1,000 population (July 2004 est.)

Death rate: 10 deaths/1,000 population (July 2004 est.)

Sex ratio:
at birth: NA
under 15 years: 1.06 male(s)/female
15-64 years: 1.01 male(s)/female
65 years and older: 0.69 male(s)/female
total population: 0.96 male(s)/female (July 2004 est.)

Infant mortality rate: *total:* 5.3 deaths/1,000 live births (July 2004 est.)

Life expectancy at birth:
total population: 78.1 years
male: 74.9 years
female: 81.4 years (July 2004 est.)

Total fertility rate: 1.48 children born/woman (July 2004 est.)

ECONOMY

Overview: Domestically, the European Union attempts to lower trade barriers, adopt a common currency, and move toward convergence of living standards. Internationally, the EU aims to bolster

Europe's trade position and its political and economic power. Because of the great differences in per capita income (from $10,000 to $28,000) and historic national animosities, the European Community faces difficulties in devising and enforcing common policies. For example, both Germany and France since 2003 have flouted the member states' treaty obligation to prevent their national budgets from running more than a 3% deficit. In 2004, the EU admitted 10 central and eastern European countries that are, in general, less advanced technologically and economically than the existing 15. The Economic and

Monetary Union (EMU), an associated organization, introduced the euro as the common currency on 1 January 1999. The UK, Sweden, and Denmark do not now participate; the 10 new countries may choose to join the EMU when they meet its fiscal and monetary criteria and the member states so agree.

GDP - per capita: purchasing power parity - $25,700 (2004 est.)

GDP - composition by sector: *agriculture:* 2.3%
industry: 28.3%
services: 69.4% (2004 est.)

Sources: The World Factbook page on European Union (2004)
http://www.odci.gov/cia/publications/factbook

Chapter References

Barzun, Jacques. 1991. *Begin Here: The Forgotten Conditions of Teaching and Learning*, edited by M. Philipson. Chicago: University of Chicago Press.

Kozol, Jonathan. 1985. *Illiterate America*. Garden City, NY: Anchor.

Lazarus, Edward. 1991. *Black Hills, White Justice: The Sioux Nation versus the United States, 1775 to the Present*. New York: HarperCollins.

Oakes, Jennie. 1985. *Keeping Track: How Schools Structure Inequality*. New Haven: Yale University Press.

Sacks, Oliver. 1989. *Seeing Voices: A Journey into the World of the Deaf*. Los Angeles: University of California Press.

Answers

Concept Application

1.	Self-fulfilling prophecy	pg. 501
2.	Formal education	pg. 481
3.	Functionally illiterate	pg. 482
4.	Tracking, ability grouping	pg. 497

Multiple-Choice

1.	c	pg. 481	13.	b	pg. 494
2.	b	pg. 480	14.	b	pg. 496
3.	c	pg. 478	15.	a	pg. 497
4.	c	pg. 481	16.	b	pg. 500
5.	a	pg. 481	17.	c	pg. 501
6.	d	pg. 482	18.	a	pg. 504
7.	c	pg. 484	19.	c	pg. 506
8.	b	pg. 487	20.	c	pg. 506
9.	c	pg. 485	21.	a	pg. 510
10.	b	pg. 487	22.	c	pg. 508
11.	c	pg. 490	23	d	pg. 515
12.	a	pg. 491	24.	c	pg. 516

True/False

1.	T	pg. 479
2.	F	pg. 485
3.	T	pg. 486
4.	T	pg. 489
5.	F	pg. 490
6.	T	pg. 491
7.	F	pg. 496
8.	T	pg. 501
9.	T	pg. 506
10.	T	pg. 499

Chapter 15

Religion
With Emphasis on the Islamic State of Afghanistan

Study Questions

1. Why was Afghanistan chosen as the country to emphasize with regard to religion?

2. When sociologists study religion, what do they study?

3. According to Durkheim, how should sociologists approach the study of religion?

4. According to Durkheim, what are three fundamental and indispensable features of religion? How do these features figure into a definition of religion?

5. Distinguish between the sacred and the profane. What are the three major types of religion, as categorized in terms of sacred phenomena?

6. According to Durkheim, what are rituals? What are the most important outcomes of rituals?

7. Distinguish between ecclesiae, denominations, sects, established sects, and cults.

8. What is civil religion? What role did civil religion play in the Cold War?

9. Why is Afghanistan's geographical location critical to its current situation?

10. What are some problems with Durkheim's definition of religion? Give examples. Are there better definitions?

11. How did Muslims come to be partners to the U.S. during the Cold War?

12. Is the question "What is religion" only of interest to sociologists? Explain.

13. What function does religion serve for the individual and the group?

14. Explain what Durkheim means by the statement, "The something out there that people worship is actually society." How is it that society is worthy of such worship?

15. Is religion strictly an integrative force? Why or why not?

16. How did Karl Marx conceptualize religion?

17. What are some criticisms of Marx's views of religion?

18. According to Weber, what role did the Protestant ethic play in the origins and development of modern capitalism? In what ways has Weber been misinterpreted?

19. What is secularization? Distinguish between Muslim views and American-European views about the causes of secularization.

20. What is fundamentalism? How are fundamentalism and secularization related?

21. What are the factors behind the surge of fundamentalism in Muslim countries?

22. Distinguish between religious and political *jihad* (including militant Islam).

23. How many militant Islamic political *jihadists* exist in the world today?

Key Concepts

Sacred	pg. 524
Sacramental religions	pg. 525
Prophetic religions	pg. 525
Mystical religions	pg. 526
Profane	pg. 526
Rituals	pg. 526
Church	pg. 527
Ecclesiae	pg. 528
Denomination	pg. 528
Sect	pg. 529
Established sects	pg. 529
Cults	pg. 532
Civil religion	pg. 532
Liberation theology	pg. 545
Modern capitalism	pg. 545
This-worldly ascetiscism	pg. 547
Predestination	pg. 548
Secularization	pg. 549
Subjective secularization	pg. 549

Fundamentalism pg. 550

Islamic revitalism pg. 551

Political *jihad* pg. 552
Religious *jihad* pg. 552

Concept Application

Consider the concepts listed below. Match one or more of the concepts with each scenario. Explain your choices.

- ✓ Church
- ✓ Civil religion
- ✓ Liberation theology
- ✓ Mystical religions
- ✓ Rituals
- ✓ Sect

Scenario 1 "As for my own religious practice, I try to live my life pursuing what I call the Bodhisattva ideal.... The Bodhisattva idea is thus the aspiration to practice infinite compassion with infinite wisdom. As a means of helping myself in the quest, I choose to be a Buddhist monk. There are 253 rules of Tibetan monasticism (364 for nuns) and by observing them as closely as I can, I free myself from many of the distractions and worries of life. Some of these rules mainly deal with etiquette, such as the physical distance a monk should walk behind the abbot of his monastery; others are concerned with behavior. The four root vows concern simple prohibitions: namely that a monk must not kill, steal, or lie about his spiritual attainment. He must also be celibate. If he breaks any one of these, he is no longer a monk" (Gyatso 1990:204-05).

Scenario 2 There were usually three services each day: morning, midafternoon, and evening. A ram's horn summoned everyone to the nine-o'clock morning service, at which time people would leave their camps and congregate in the shed. Sunday was the biggest day of the week, and for many years it was also the day of the Lovefeast. Bread and water were passed around and people would make their testimonials. During the evening service there would inevitably be an altar call, often accompanied by a lot of shouting (Jenkins 1996:562).

Scenario 3 "The 'miracle' was Brazil's accelerated economic growth between 1968 and 1975; Brazil moved from twenty-first to fourteenth in rank among developing countries, based upon per capita GNP. The 'miracle' did not help most Brazilians, however. The imbalances in the distribution of wealth were made yet worse. The Brazilian bishops have openly denounced the 'Brazilian miracle' for the poverty it has engendered. They have attacked the economic policies that have pushed thousands of peasant farmers off the lands their families have farmed for generations, and they have questioned development projects (such as the exploitation of the Amazon) which displaced the native Indians and poor farmers but brought them no benefit. Indeed, one observer has concluded that 'the church has become the primary institutional focus of dissidence in the country'" (McGuire 1987:215).

Scenario 4 "Mennonites trace their roots to a small group of Christians after 1530 who sought a reformation even more radical than those advocated by Lutherans and Calvinists. They were called Mennonites after Menno Simons, one of their early leaders. Their most distinctive practice is adult baptism offered only to those who have made a decision to follow Christ's teachings" (Lorimer 1989:212).

Scenario 5 The old fascist marching songs were sung, a moment of silence was observed for all who died defending the fatherland, and the gathering was reminded that today was the 57th anniversary of the founding of Croatia's Nazi-allied wartime government. Then came the most chilling words of the afternoon.

"For Home!" shouted Anto Dapic, surrounded by bodyguards in black suits and crew cuts.

"Ready!" responded the crowd of 500 supporters, their arms rising in a stiff Nazi salute.

The call and response—the Croatian equivalent of "Sieg!" "Heil!"—was the wartime greeting used by supporters of the fascists Independent Sate of Croatia that governed the country for most the Second World War and murdered hundreds of thousands of Jews, Serbs and Croatian resistance fighters (Hedges 1997:3Y).

Applied Research

If you are a member of an organized religion, visit a church of another denomination. Choose two sociological concepts from Chapter 15. Write a paper comparing and contrasting the two services or religious events.

Practice Test: Multiple-Choice Questions

1. In studying the Taliban, sociologists would be least likely to ask which one of the following questions?
   a. How did the Taliban rise to power?
   b. Is the Taliban's version of Islamic law consistent with Islamic principles?
   c. How did religious fanatics acting solely from "primitive and irrational religious convictions" gain so much power?
   d. Why has Afghanistan experienced civil war for the past 20 years?

2. __________ wrote "To define 'religion,' to say what it is, is not possible at the start of a presentation such as this. Definition can be attempted, if at all, only at the conclusion of the study."
   a. Karl Marx
   b. Max Weber
   c. Emile Durkheim
   d. George H. Mead

3. Sacred things can include books, buildings, days, and places. From a sociological point of view sacredness stems from
   a. the item itself.
   b. an item's symbolic power.
   c. the meaning assigned to it by God.
   d. thc bible.

4. Which of the following statements does not apply to Native Spirituality?
   a. There are probably as many native religions as there are Indian tribes.
   b. The basic tenets of Native Spirituality can be found in the "Great Book".
   c. None of the native religions have man-made churches in the Judeo-Christian sense.
   d. Religious beliefs are tied to nature.

5. Some of the most well-known _________ religions include Judaism, Confucianism, Christianity, and Islam.
   a. sacramental
   b. prophetic
   c. mystical
   d. profane

6. Confession, immersion, and fasting are examples of
   a. mystical acts.
   b. ecclesiae.
   c. rituals.
   d. sacraments.

7. __________ include(s) everything that is not sacred.
   a. Powerful symbols
   b. Evil
   c. The profane
   d. Exorcism

8. Durkheim used the word "church" to designate a group whose members do all but which one of the following?
   a. hold the same beliefs with regard to the sacred and the profane
   b. behave in the same way in the presence of the sacred
   c. gather together to affirm commitment to beliefs and practices
   d. adhere to the belief that the religion members follow is one of many true religions

9. In Islam the most pronounced split occurred after the death of Prophet Muhammad over the issue of Muhammed's successor. That split is between
   a. Sunni and Shia.
   b. Hezbollah and Druze.
   c. Iranian Sunni and Iraqi Shia.
   d. Muslims and Jews.

10. ________________ is the ancient native religion of Japan.
    a. Judaism
    b. Taoism
    c. Islam
    d. Shinto

11. In light of Durkheim's definition of religion, which one of the following does not qualify as a religious phenomenon?
    a. displays of patriotism
    b. 21-gun salutes
    c. national holidays
    d. traffic jams in which everyone gets out of their cars to interact

12. When the Soviet Union invaded Afghanistan in 1979, the U.S.
    a. let the Soviets control the country.
    b. dropped an atomic bomb.
    c. sent 250,000 troops to the region.
    d. supported the Afghan freedom fighters, known as the *mujahidin*.

13. President George W. Bush's Faith Based and Community Initiative is an extension of
    a. Liberation theology.
    b. the Department of Homeland Security.
    c. post 9-11 initiatives.
    d. Charitable Choice, a Clinton Administration initiative.

14. The Soviet Union invaded Afghanistan in
    a. 1979.
    b. 1994.
    c. 1945.
    d. 1890.

15. The United States worked with _____________ to recruit 35,000 Muslims from 43 countries to fight with their Afghan brothers against the Soviet Union.
    a. Iran
    b. Pakistan
    c. Tajikistan
    d. India

16. Functionalists maintain that religion must serve some vital social function because
    a. there are very few atheists in the world.
    b. all people turn to religion in times of deep distress.
    c. some form of religion has existed as long as humans have been around.
    d. people who communicate with their god find extraordinary strength.

17. Which one of the following sociologists observed that whenever a group of people have a strong conviction, that conviction almost always takes on a religious character?
    a. Robert Coles
    b. Robert K. Merton
    c. Emile Durkheim
    d. Max Weber

18. If religion were truly an integrative force
    a. there would be no conflict or tensions among religious groups within the same society.
    b. everyone would have the same religion.
    c. there would be fewer struggles between the political and the religious.
    d. everyone would be a member of a religion.

19. Durkheim observed that whenever a group of people has a strong conviction
    a. religious values are secondary to the conviction.
    b. that conviction always takes on a religious character.
    c. they fight among themselves.
    d. they work to make the world a better place.

20. __________ is/are an example of a religion that emerged in the United States in the 1930s as a vehicle of protest or change.
    a. Liberation theology
    b. The Quakers
    c. Black Shia
    d. Nation of Islam

21. Danielle believes that God has foreordained all things including the salvation or damnation of individual souls. This belief is known as
    a. liberation theology.
    b. secularization.
    c. predestination.
    d. fundamentalism.

22. Weber maintained that the Protestant ethic
    a. caused capitalism to come into being.
    b. led to the rise of fundamentalism.
    c. was a significant force in the emergence of capitalism.
    d. must be present in a society if it is to achieve economic success.

23. The phrase "under God" was inserted into the U.S. Pledge of Allegiance in the
    a. 1750s.
    b. 1880s.
    c. 1930s.
    d. 1950s.

24. Daniel Pipes estimates that there are ______________ million persons "who do not accept the particulars" of militant Islam but are sympathetic and supportive of the anti-American stance.
    a. 1
    b. 10
    c. 100
    d. 500

True/False Questions

T F 1. The Taliban rose to power in Afghanistan after the September 11 attacks on the United States.

T F 2. In studying religions, sociologists must assume that there are no religions that are false.

T F 3. Rituals can be codes of conduct aimed at governing the performance of everyday activities such as eating.

T F 4. All the major religions encompass splinter groups that have sought to preserve the integrity of their religion.

T F 5. Cults often dissolve after their leader dies.

T F 6. Durkheim maintained that whenever a group of people has a strong conviction, it almost always takes on a religious character.

T F 7. Karl Marx maintained that people need the comfort of religion in order to make the world bearable and justify their existence in it.

T F 8. The Protestant ethic <u>caused</u> capitalism to emerge.

T F 9. Fundamentalism is a process by which religious influences on thought and behavior are reduced.

T F 10. Archeological evidence suggests that Jesus was at least 6 feet 2 inches tall.

Internet Resources Related to Religion

- **About Specific Religions, Faith Groups, Ethical Systems, etc.**
  http://www.religioustolerance.org/var_rel.htm
  This website offers information about religions, faith groups and ethical systems. They are broken down into the following categories: 1) World Religions – long established religions such as Buddhism, Judaism and Christianity; 2) Neopagan Religious Faiths – "modern-day reconstructions of ancient pagan religions" such as witchcraft and Druidism; 3) Other organized Religions – "well defined belief in deity, humanity and the rest of the universe" such as Scientology and Unitarian-Universalism; 4) Destructive Doomsday Cults – "religiously based, very high intensity, controlling groups that have caused or are liable to cause loss of life among their membership or the general public" such as Branch Dividian and Heaven's Gate (note: these are located in the "Introductory Thoughts" link.) This website is supported by the Ontario Consultants on Religious Tolerance whose aims are to promote tolerance of minority religions, offer useful information on controversial religious topics, and expose hatred and misinformation about any religion.

Internet Resources Related to Afghanistan

- **Afghanistan News.net**
  http://afghanistannews.net/
  For the latest news on Afghanistan visit this website.

- **Afghanistan's Website**
  http://www.afghanistans.com/
  Learn about Afghanistan's land and resources, the people, the climate, past and present flags, proverbs, alphabet, music, view photos and daily news coverage.

Afghanistan: Background Notes

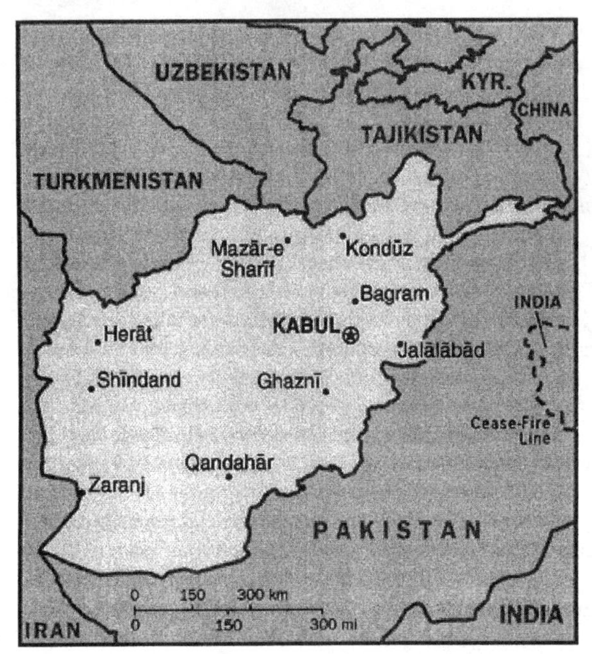

PEOPLE

Nationality: *Noun and adjective*--Afghan(s).
Population: 28,717,213 (July 2003 est.). More than 4 million Afghans live outside the country, mainly in Pakistan and Iran, although over two and a half million have returned since the removal of the Taliban. Annual population growth rate (2003 est.): 3.38%. This rate does not take into consideration the recent war and its continuing impact.
Main ethnic groups: Pashtun, Tajik, Hazara, Uzbek, Turkmen, Aimaq, Baluch, Nuristani, Kizilbash.
Religions: Sunni Muslim 84%, Shi'a Muslim 15%, other 1%.
Main languages: Dari (Afghan Persian), Pashto.
Education: Approximately 4 million children, of whom some 30% are girls, enrolled in school during 2003. *Literacy* (2001 est.)--36% (male 51%, female 21%), but real figures may be lower given breakdown of education system and flight of educated Afghans.
Health: *Infant mortality rate* (2003)—142.48/1,000. *Life expectancy* (2003 est.)--47.67 yrs. (male); 46.23 yrs. (female).
Work force: Mostly in rural agriculture; number cannot be estimated due to conflict.

Afghanistan's ethnically and linguistically mixed population reflects its location astride historic trade and invasion routes leading from Central Asia into South and Southwest Asia. Pashtuns are the dominant ethnic group, accounting for about 38-44% of the population. Tajik (25%), Hazara (10-19%), Uzbek (6-8%), Aimaq, Turkmen, Baluch, and other small groups also are represented. Dari (Afghan Persian) and Pashto are official languages. Dari is spoken by more than one-third of the population as a first language and serves as a lingua franca for most Afghans, though the Taliban use Pashto. Tajik, Uzbek, and Turkmen are spoken widely in the north. Smaller groups throughout the country also speak more than 70 other languages and numerous dialects.

Afghanistan is an Islamic country. An estimated 84% of the population is Sunni, following the Hanafi school of jurisprudence; the remainder is predominantly Shi'a, mainly Hazara. Despite attempts during the years of communist rule to secularize Afghan society, Islamic practices pervade all aspects of life. In fact, Islam served as the principal basis for expressing opposition to the communists and the Soviet invasion. Likewise, Islamic religious tradition and codes, together with traditional practices, provide the principal means of controlling personal conduct and settling legal disputes. Excluding urban populations in the principal cities, most Afghans are divided into tribal and other kinship-based groups, which follow traditional customs and religious practices.

HISTORY

Afghanistan, often called the crossroads of Central Asia, has had a turbulent history. In 328 BC, Alexander the Great entered the territory of present-day Afghanistan, then part of the Persian Empire, to capture Bactria (present-day Balkh). Invasions by the Scythians, White Huns, and Turks followed in succeeding centuries. In AD 642, Arabs invaded the entire region and introduced Islam.

Arab rule gave way to the Persians, who controlled the area until conquered by the Turkic Ghaznavids in 998. Mahmud of Ghazni (998-1030) consolidated the conquests of his predecessors and turned Ghazni into a great cultural center as well as a base for frequent forays into India. Following Mahmud's short-lived dynasty, various princes attempted to rule sections of the country until the Mongol invasion of 1219. The Mongol invasion, led by Genghis Khan, resulted in massive slaughter of the population, destruction of many cities, including Herat, Ghazni, and Balkh, and the despoliation of fertile agricultural areas.

Following Genghis Khan's death in 1227, a succession of petty chiefs and princes struggled for supremacy until late in the 14th century, when one of his descendants, Tamerlane, incorporated Afghanistan into his own vast Asian empire. Babur, a descendant of Tamerlane and the founder of India's Moghul dynasty at the beginning of the 16th century, made Kabul the capital of an Afghan principality.

In 1747, Ahmad Shah Durrani, the founder of what is known today as Afghanistan, established his rule. A Pashtun, Durrani was elected king by a tribal council after the assassination of the Persian ruler Nadir Shah at Khabushan in the same year. Throughout his reign, Durrani consolidated chieftainships, petty principalities, and fragmented provinces into one country. His rule extended from Mashad in the west to Kashmir and Delhi in the east, and from the Amu Darya (Oxus) River in the north to the Arabian Sea in the south. With the exception of a 9-month period in 1929, all of Afghanistan's rulers until the 1978 Marxist coup were from Durrani's Pashtun tribal confederation, and all were members of that tribe's Mohammadzai clan after 1818.

European Influence

During the 19th century, collision between the expanding British Empire in the subcontinent and czarist Russia significantly influenced Afghanistan in what was termed "The Great Game." British concern over Russian advances in Central Asia and growing influence in Persia culminated in two Anglo-Afghan wars. The first (1839-42) resulted not only in the destruction of a British army, but is remembered today as an example of the ferocity of Afghan resistance to foreign rule. The second Anglo-Afghan war (1878-80) was sparked by Amir Sher Ali's refusal to accept a British mission in Kabul. This conflict brought Amir Abdur Rahman to the Afghan throne. During his reign (1880-1901), the British and Russians officially established the boundaries of what would become modern Afghanistan. The British retained effective control over Kabul's foreign affairs.

Afghanistan remained neutral during World War I, despite German encouragement of anti-British feelings and Afghan rebellion along the borders of British India. The Afghan king's policy of neutrality was not universally popular within the country, however.

Habibullah, Abdur Rahman's son and successor, was assassinated in 1919, possibly by family members opposed to British influence. His third son, Amanullah, regained control of Afghanistan's foreign policy after launching the third Anglo-Afghan war with an attack on India in the same year. During the ensuing conflict, the war-weary British relinquished their control over Afghan foreign affairs by signing the Treaty of Rawalpindi in August 1919. In commemoration of this event, Afghans celebrate August 19 as their Independence Day.

Reform and Reaction

King Amanullah (1919-29) moved to end his country's traditional isolation in the years following the third Anglo-Afghan war. He established diplomatic relations with most major countries and, following a 1927 tour of Europe and Turkey--during which he noted the modernization and secularization advanced by Ataturk--introduced several reforms intended to modernize Afghanistan. Some of these, such as the abolition of the traditional Muslim veil for women and the opening of a number of co-educational schools, quickly alienated many tribal and religious leaders. Faced with overwhelming armed opposition, Amanullah was forced to abdicate in January 1929 after Kabul fell to forces led by Bacha-i-Saqao, a Tajik brigand. Prince Nadir Khan, a cousin of Amanullah's, in turn defeated Bacha-i-Saqao in October of the same year and, with considerable Pashtun tribal support, was declared King Nadir Shah. Four years later, however, he was assassinated in a revenge killing by a Kabul student.

Mohammad Zahir Shah, Nadir Khan's 19-year-old son, succeeded to the throne and reigned from 1933 to 1973. In 1964, King Zahir Shah promulgated a liberal constitution providing for a two-chamber legislature to which the king appointed one-third of the deputies. The people elected another third, and the remainder were selected indirectly by provincial assemblies. Although Zahir's "experiment in democracy" produced few lasting reforms, it permitted the growth of unofficial extremist parties on both the left and the right. These included the communist People's Democratic Party of Afghanistan (PDPA), which had close ideological ties to the Soviet Union. In 1967, the PDPA split into two major rival factions: the Khalq (Masses) faction headed by Nur Muhammad Taraki and Hafizullah Amin and supported by elements within the military, and the Parcham (Banner) faction led by Babrak Karmal. The split reflected ethnic, class, and ideological divisions within Afghan society.

Zahir's cousin, Sardar Mohammad Daoud, served as his Prime Minister from 1953 to 1963. During his tenure as Prime Minister, Daoud solicited military and economic assistance from both Washington and Moscow and introduced controversial social policies of a reformist nature. Daoud's alleged support for the creation of a Pashtun state in the Pakistan-Afghan border area heightened tensions with Pakistan and eventually resulted in Daoud's dismissal in March 1963.

Daoud's Republic (1973-78) and the April 1978 Coup

Amid charges of corruption and malfeasance against the royal family and poor economic conditions created by the severe 1971-72 drought, former Prime Minister Daoud seized power in a military coup on July 17, 1973. Zahir Shah fled the country, eventually finding refuge in Italy. Daoud abolished the monarchy, abrogated the 1964 constitution, and declared Afghanistan a republic with himself as its first President and Prime Minister. His attempts to carry out badly needed economic and social reforms met with little success, and the new constitution promulgated in February 1977 failed to quell chronic political instability.

Seeking to exploit more effectively mounting popular disaffection, the PDPA reunified with Moscow's support. On April 27, 1978, the PDPA initiated a bloody coup, which resulted in the overthrow and murder of Daoud and most of his family. Nur Muhammad Taraki, Secretary General of the PDPA, became President of the Revolutionary Council and Prime Minister of the newly established Democratic Republic of Afghanistan.

Opposition to the Marxist government emerged almost immediately. During its first 18 months of rule, the PDPA brutally imposed a Marxist-style "reform" program, which ran counter to deeply rooted Afghan traditions. Decrees forcing changes in marriage customs and pushing through an ill-conceived land reform were particularly misunderstood by virtually all Afghans. In addition, thousands of members of the traditional elite, the religious establishment, and the intelligentsia were imprisoned, tortured, or murdered. Conflicts within the PDPA also surfaced early and resulted in exiles, purges, imprisonments, and executions.

By the summer of 1978, a revolt began in the Nuristan region of eastern Afghanistan and quickly spread into a countrywide insurgency. In September 1979, Hafizullah Amin, who had earlier been Prime Minister and Minister of Defense, seized power from Taraki after a palace shootout. Over the next 2 months, instability plagued Amin's regime as he moved against perceived enemies in the PDPA. By December, party morale was crumbling, and the insurgency was growing.

The Soviet Invasion

The Soviet Union moved quickly to take advantage of the April 1978 coup. In December 1978, Moscow signed a new bilateral treaty of friendship and cooperation with Afghanistan, and the Soviet military assistance program increased significantly. The regime's survival increasingly was dependent upon Soviet military equipment and advisers as the insurgency spread and the Afghan army began to collapse.

By October 1979, however, relations between Afghanistan and the Soviet Union were tense as Hafizullah Amin refused to take Soviet advice on how to stabilize and consolidate his government. Faced with a deteriorating security situation, on December 24, 1979, large numbers of Soviet airborne forces, joining thousands of Soviet troops already on the ground, began to land in Kabul under the pretext of a field exercise. On December 26, these invasion forces killed Hafizullah Amin and installed Babrak Karmal, exiled leader of the Parcham faction, bringing him back from Czechoslovakia and making him Prime Minister. Massive Soviet ground forces invaded from the north on December 27.

Following the invasion, the Karmal regime, although backed by an expeditionary force that grew as large as 120,000 Soviet troops, was unable to establish authority outside Kabul. As much as 80% of the countryside, including parts of Herat and Kandahar, eluded effective government control. An overwhelming majority of Afghans opposed the communist regime, either actively or passively. Afghan freedom fighters (mujahidin) made it almost impossible for the regime to maintain a system of local government outside major urban centers. Poorly armed at first, in 1984 the mujahidin began receiving substantial assistance in the form of weapons and training from the U.S. and other outside powers.

In May 1985, the seven principal Peshawar-based guerrilla organizations formed an alliance to coordinate their political and military operations against the Soviet occupation. Late in 1985, the mujahidin were active in and around Kabul, launching rocket attacks and conducting operations against the communist government. The failure of the Soviet Union to win over a significant number of Afghan collaborators or to rebuild a viable Afghan army forced it to bear an increasing responsibility for fighting the resistance and for civilian administration.

Soviet and popular displeasure with the Karmal regime led to its demise in May 1986. Karmal was replaced by Muhammad Najibullah, former chief of the Afghan secret police (KHAD). Najibullah had established a reputation for brutal efficiency during his tenure as KHAD chief. As Prime Minister, Najibullah was ineffective and highly dependent on Soviet support. Undercut by deep-seated divisions within the PDPA, regime efforts to broaden its base of support proved futile.

The Geneva Accords and Their Aftermath

By the mid-1980s, the tenacious Afghan resistance movement--aided by the United States, Saudi Arabia, Pakistan, and others--was exacting a high price from the Soviets, both militarily within Afghanistan and by souring the U.S.S.R.'s relations with much of the Western and Islamic world. Informal negotiations for a Soviet withdrawal from Afghanistan had been underway since 1982. In 1988, the Governments of Pakistan and Afghanistan, with the United States and Soviet Union serving as guarantors, signed an agreement settling the major differences between them. The agreement, known as the Geneva accords, included five major documents, which, among other things, called for U.S. and Soviet noninterference in the internal affairs of Pakistan and Afghanistan, the right of refugees to return to Afghanistan without fear of persecution or harassment, and, most importantly, a timetable that ensured full Soviet withdrawal from Afghanistan by February 15, 1989. About 14,500 Soviet and an estimated one million Afghan lives were lost between 1979 and the Soviet withdrawal in 1989.

Significantly, the mujahidin were party neither to the negotiations nor to the 1988 agreement and, consequently, refused to accept the terms of the accords. As a result, the civil war continued after the Soviet withdrawal, which was completed in February 1989. Najibullah's regime, though failing to win popular support, territory, or international recognition, was able to remain in power until 1992 but collapsed after the defection of Gen. Abdul Rashid Dostam and his Uzbek militia in March. However, when the victorious mujahidin entered Kabul to assume control over the city and the central government, a new round of internecine fighting began between the various militias, which had coexisted only uneasily during the Soviet occupation. With the demise of their common enemy, the militias' ethnic, clan, religious, and personality differences surfaced, and the civil war continued.

Seeking to resolve these differences, the leaders of the Peshawar-based mujahidin groups established an interim Islamic Jihad Council in mid-April 1992 to assume power in Kabul. Moderate leader Prof. Sibghatullah Mojaddedi was to chair the council for 2 months, after which a 10-member leadership council composed of mujahidin leaders and presided over by the head of the Jamiat-i-Islami, Prof. Burhanuddin Rabbani, was to be set up for 4 months. During this 6-month period, a Loya Jirga, or grand council of Afghan elders and notables, would convene and designate an interim administration which would hold power up to a year, pending elections.

But in May 1992, Rabbani prematurely formed the leadership council, undermining Mojaddedi's fragile authority. In June, Mojaddedi surrendered power to the Leadership Council, which then elected Rabbani as President. Nonetheless, heavy fighting broke out in August 1992 in Kabul between forces loyal to President Rabbani and rival factions, particularly those who supported Gulbuddin Hekmatyar's Hezb-i-Islami. After Rabbani extended his tenure in December 1992, fighting in the capital flared up in January and February 1993. The Islamabad Accord, signed in March 1993, which appointed Hekmatyar as Prime Minister, failed to have a lasting effect. A follow-up agreement, the Jalalabad Accord, called for the militias to be disarmed but was never fully implemented. Through 1993, Hekmatyar's Hezb-i-Islami forces, allied with the Shi'a Hezb-i-Wahdat militia, clashed intermittently with Rabbani and Masood's Jamiat forces. Cooperating with Jamiat were militants of Sayyaf's Ittehad-i-Islami and, periodically, troops loyal to ethnic Uzbek strongman Abdul Rashid Dostam. On January 1, 1994, Dostam switched sides, precipitating large-scale fighting in Kabul and in northern provinces, which caused thousands of civilian casualties in Kabul and elsewhere and created a new wave of displaced persons and refugees. The country sank even further into anarchy, forces loyal to Rabbani and Masood, both ethnic Tajiks, controlled Kabul and much of the northeast, while local warlords exerted power over the rest of the country.

Rise of the Taliban

In reaction to the anarchy and warlordism prevalent in the country, and the lack of Pashtun representation in the Kabul government, a movement of former mujahidin arose. Many Taliban had been educated in madrassas in Pakistan and were largely from rural Pashtun backgrounds. The name "Talib" itself means pupil. This group dedicated itself to removing the warlords, providing order, and imposing Islam on the country. It received considerable support from Pakistan. In 1994, it developed enough strength to capture the city of Kandahar from a local warlord and proceeded to expand its control throughout Afghanistan, occupying Kabul in September 1996. By the end of 1998, the Taliban occupied about 90% of the country, limiting the opposition largely to a small mostly Tajik corner in the northeast and the Panjshir valley. Efforts by the UN, prominent Afghans living outside the country, and other interested countries to bring about a peaceful solution to the continuing conflict came to naught, largely because of intransigence on the part of the Taliban.

The Taliban sought to impose an extreme interpretation of Islam--based in part upon rural Pashtun tradition--on the entire country and committed massive human rights violations, particularly directed against women and girls, in the process. Women were restricted from working outside the home and pursuing an education, were not to leave their homes without an accompanying male relative, and were forced to wear a traditional body-covering garment called the burka. The Taliban committed serious atrocities against minority populations, particularly the Shi'a Hazara ethnic group, and killed noncombatants in several well-documented instances. In 2001, as part of a drive against relics of Afghanistan's pre-Islamic past, the Taliban destroyed two large statues of the Buddha outside of the city of Bamiyan and announced destruction of all pre-Islamic statues in Afghanistan, including the remaining holdings of the Kabul Museum.

From the mid-1990s the Taliban provided sanctuary to Osama bin Laden, a Saudi national who had fought with them against the Soviets, and provided a base for his and other terrorist organizations. The UN Security Council repeatedly sanctioned the Taliban for these activities. Bin Laden provided both financial and political support to the Taliban. Bin Laden and his al Qaeda group were charged with the bombing of the U.S. Embassies in Nairobi and Dar Es Salaam in 1998, and in August 1998 the United States launched a cruise missile attack against bin Laden's terrorist camp in Afghanistan. Bin Laden and al Qaeda are believed to be responsible for the September 11, 2001 terrorist acts in the United States, among other crimes.

In September 2001, agents working on behalf of the Taliban and believed to be associated with bin Laden's al Qaeda group assassinated Northern Alliance Defense Minister and chief military commander Ahmed Shah Masood, a hero of the Afghan resistance against the Soviets and the Taliban's principal military opponent. Following the Taliban's repeated refusal to expel bin Laden and his group and end its support for international terrorism, the U.S. and its partners in the anti-terrorist coalition began a campaign on October 7, 2001, targeting terrorist facilities and various Taliban military and political assets within Afghanistan.

Under pressure from U.S. air power and anti-Taliban ground forces, the Taliban disintegrated rapidly, and Kabul fell on November 13, 2001. Sponsored by the UN, Afghan factions opposed to the Taliban met in Bonn, Germany in early December and agreed to restore stability and governance to Afghanistan by creating an interim government and establishing a process to move toward a permanent government. Under this so-called Bonn Agreement, an Afghan Interim Authority was formed and took office in Kabul on December 22, 2001 with Hamid Karzai as Chairman. The Interim Authority held power for approximately 6 months while preparing for a nationwide "Loya Jirga" (Grand Council) in mid-June 2002 that decided on the structure of a Transitional Authority. The Transitional Authority, headed by President Hamid Karzai, renamed the government as the Transitional Islamic State of Afghanistan (TISA). One of the TISA's primary achievements was the drafting of a constitution that was ratified by a Constitutional Loya Jirga on January 4, 2004.

ECONOMY

GDP: $4 billion (2002-03 est.).
Per capita GDP: $180-$190 (based on 22 million population estimate).
Purchasing parity power: $19 billion (2002 est.)
GDP growth: 28.6% (2002-03 est.)
Natural resources: Natural gas, oil, coal, copper, chromite, talc, barites, sulfur, lead, zinc, iron, salt, precious and semiprecious stones.

Agriculture (estimated 52% of GDP): *Products*--wheat, corn, barley, rice, cotton, fruit, nuts, karakul pelts, wool, and mutton.
Industry (estimated 26% of GDP): *Types*--small-scale production for domestic use of textiles, soap, furniture, shoes, fertilizer, and cement; hand-woven carpets for export; natural gas, precious and semiprecious gemstones.
Services (estimated 22% of GDP): transport, retail, and telecommunications.
Trade (2002-03 est.): *Exports*--$100 million (does not include opium): fruits and nuts, handwoven carpets, wool, cotton, hides and pelts, precious and semiprecious gems. *Major markets*--Central Asian republics, Pakistan, Iran, EU, India. Estimates show that the figure for 2001 was much lower, except for opium. *Imports*--$2.3 billion: food, petroleum products, machinery, and consumer goods. Estimates show that imports were severely reduced in 2001. *Major suppliers*--Central Asian republics, Pakistan, Iran.
Currency: The currency is the afghani, which was reintroduced as Afghanistan's new currency in January 2003. The exchange rate of the new currency has remained broadly stable since the completion of the conversion process from the country's old afghani currency. At present, $1 U.S. equals approximately 43 afghanis. Since its inception the new afghani has gained gradual acceptance throughout the country, but other foreign currencies are also still frequently accepted as legal tender.

In the 1930s, Afghanistan embarked on a modest economic development program. The government founded banks; introduced paper money; established a university; expanded primary, secondary, and technical schools; and sent students abroad for education. In 1956, the Afghan Government promulgated the first in a long series of ambitious development plans. By the late 1970s, these had achieved only mixed results due to flaws in the planning process as well as inadequate funding and a shortage of the skilled managers and technicians needed for implementation.

Historically, there has been a dearth of information and reliable statistics about Afghanistan's economy. The 1979 Soviet invasion and ensuing civil war destroyed much of the underdeveloped country's limited infrastructure and disrupted normal patterns of economic activity. Gross domestic product had fallen substantially over the preceding 23 years because of loss of labor and capital and disruption of trade and transport. Continuing internal strife hampered both domestic efforts at reconstruction as well as international aid efforts. However, Afghanistan's economy has been growing at a fast pace since the 2001 fall of the Taliban, albeit from a low base. In 2003, growth was estimated at close to 30%, and the growth rate is expected to be over 20% in 2004.

Agriculture
The Afghan economy continues to be overwhelmingly agricultural, despite the fact that only 12% of its total land area is arable and less than 6% currently is cultivated. Agricultural production is constrained by an almost total dependence on erratic winter snows and spring rains for water; irrigation is primitive. Relatively little use is made of machines, chemical fertilizer, or pesticides.

Grain production is Afghanistan's traditional agricultural mainstay. Overall agricultural production dramatically declined following 4 years of severe drought as well as sustained fighting, instability in rural areas, and deteriorated infrastructure. Soviet efforts to disrupt production in resistance-dominated areas also contributed to this decline, as did the disruption to transportation resulting from ongoing conflict. The easing of the drought, which had affected more than half of the population into late 2002, and the end of civil war produced the largest wheat harvest in 25 years during 2003. Wheat production was an estimated 58% higher than in 2002. However, the country still needed to import an estimated million tons of wheat to meet its requirements for the year. Millions of Afghans, particularly in rural areas, remained dependent on food aid.

The war against the Soviet Union and the ensuing civil war led to migration to the cities and refugee flight to Pakistan and Iran, further disrupting normal agricultural production. Shortages were exacerbated by the country's already limited transportation network, which had deteriorated further due to damage and neglect resulting from war and the absence of an effective central government. Agricultural production and livestock numbers are still not sufficient to feed a large percentage of Afghanistan's population.

Opium has became a source of cash for many Afghans, especially following the breakdown in central authority after the Soviet withdrawal, and opium-derived revenues probably constituted a major source of income for the two main factions during the civil war in the 1990s. The Taliban earned roughly $40 million per year on opium taxes alone. Opium is easy to cultivate and transport and offers a quick source of income for impoverished Afghans. Afghanistan was the world's largest producer of raw opium in 1999 and 2000. Much of Afghanistan's opium production is refined into heroin and is either consumed by a growing regional addict population or exported, primarily to Western Europe. Despite efforts to bring opium cultivation under control, the most recent 2003 crop is reportedly the largest recorded. The international community and the new Afghan Government are currently working on new initiatives to eliminate the narcotics economy.

Trade and Industry

Trade accounts for a small portion of the documented Afghan economy, and there are no reliable statistics relating to trade flows. In 2002-03, exports--not including opium or re-exports--were estimated at $100 million and imports estimated at $2.3 billion, a significant increase over 2001-02. Since the 1989 Soviet withdrawal and the 1991 collapse of the Soviet Union, other limited trade relationships with Central Asian states appear to be emerging. Exports to Iran and Pakistan account for about onc-half of total exports. Belgium, Russia, Germany, the United Arab Emirates, and the United States each account for 5% or more of Afghanistan's exports. Japan, Korea, and Pakistan account for about 40% of imports. Other significant sources of imports are Germany, India, Iran, Kenya, Turkmenistan, and the United States. While the United States revoked Afghanistan's most-favored-nation (MFN) trading status in 1986, it reestablished normal trade relations in June 2002. Most of Afghanistan's exports (excluding illegal or smuggled exports) are agricultural products and carpets.

Afghanistan is endowed with a wealth of natural resources, including extensive deposits of natural gas, petroleum, coal, copper, chromite, talc, barites, sulfur, lead, zinc, iron ore, salt, and precious and semiprecious stones. In the 1970s the Soviets estimated Afghanistan had as much as five trillion cubic feet (tcf) of natural gas, 95 million barrels of oil and condensate reserves, and 400 million tons of coal. Unfortunately, ongoing instability in certain areas of the country, remote and rugged terrain, and inadequate infrastructure and transportation network have made mining these resources difficult, and there have been few serious attempts to further explore or exploit them.

The most important resource has been natural gas, first tapped in 1967. At their peak during the 1980s, natural gas sales accounted for $300 million a year in export revenues (56% of the total). Ninety percent of these exports went to the Soviet Union to pay for imports and debts. However, during the withdrawal of Soviet troops in 1989, Afghanistan's natural gas fields were capped to prevent sabotage by the mujahidin. Restoration of gas production has been hampered by internal strife and the disruption of traditional trading relationships following the collapse of the Soviet Union. Gas production dropped from a high of 290 million cubic feet (Mmcf) per day in the 1980s to a low of about 22 Mmcf in 2001.

Trade in goods smuggled into Pakistan once constituted a major source of revenue for Afghan regimes, including the Taliban, and still figures as an important element in the Afghan economy. Many of the goods smuggled into Pakistan originally entered Afghanistan from Pakistan, where they fell under the Afghan Trade and Transit Agreement (ATTA), which permitted goods bound for Afghanistan to transit

Pakistan free of duty. When Pakistan clamped down in 2000 on the types of goods permitted duty-free transit, routing of goods through Iran from the Gulf increased significantly. Shipments of smuggled goods were subjected to fees and duties paid to the Afghan Government. The trade also provided jobs to tens of thousands of Afghans on both sides of the Durand Line, which forms the border between Afghanistan and Pakistan. Pakistan's closing its Afghan border in September 2001 presumably curtailed this traffic.

Transportation
Landlocked Afghanistan has no functioning railways, but the Amu Darya (Oxus) River, which forms part of Afghanistan's border with Turkmenistan, Uzbekistan, and Tajikistan, has barge traffic. During their occupation of the country, the Soviets completed a bridge across the Amu Darya and built a motor vehicle and railroad bridge between Termez and Jeyretan. The U.S., in conjunction with the governments of Afghanistan and Tajikistan, is currently exploring the feasibility of resuscitating a bridge link over the Amu Darya.

Most road building occurred in the 1960s, funded by the U.S. and the Soviet Union. The Soviets built a road and tunnel through the Salang Pass in 1964, connecting northern and southern Afghanistan. A highway connecting the principal cities of Herat, Kandahar, Ghazni, and Kabul with links to highways in neighboring Pakistan formed the primary road system.

Afghanistan's national airline, Ariana, operates domestic and international routes, including flights to New Delhi, Islamabad, Dubai, Moscow, Istanbul, Tehran, and Frankfurt. A private carrier, Kam Air, commenced domestic operations in November 2003.

Many sections of Afghanistan's highway and regional road system are undergoing significant reconstruction. The U.S. (with assistance from Japan) completed building a highway linking Kabul to the southern regional capital, Kandahar. Construction is soon to begin on the next phase of highway reconstruction between Kandahar and the western city of Herat. The Asian Development Bank is nearing completion on a road reconstruction project between Kandahar and Spin Boldak, located at the southeastern border with Pakistan.

U.S.-AFGHAN RELATIONS

The first extensive American contact with Afghanistan was made by Josiah Harlan, an adventurer from Pennsylvania who was an adviser in Afghan politics in the 1830s and reputedly inspired Rudyard Kipling's story "The Man Who Would be King." After the establishment of diplomatic relations in 1934, the U.S. policy of helping developing nations raise their standard of living was an important factor in maintaining and improving U.S.-Afghan ties. From 1950 to 1979, U.S. foreign assistance provided Afghanistan with more than $500 million in loans, grants, and surplus agricultural commodities to develop transportation facilities, increase agricultural production, expand the educational system, stimulate industry, and improve government administration.

In the 1950s, the U.S. declined Afghanistan's request for defense cooperation but extended an economic assistance program focused on the development of Afghanistan's physical infrastructure--roads, dams, and power plants. Later, U.S. aid shifted from infrastructure projects to technical assistance programs to help develop the skills needed to build a modern economy. The Peace Corps was active in Afghanistan between 1962 and 1979.

After the April 1978 coup, relations deteriorated. In February 1979, U.S. Ambassador Adolph "Spike" Dubs was murdered in Kabul after Afghan security forces burst in on his kidnapers. The U.S. then reduced bilateral assistance and terminated a small military training program. All remaining assistance agreements were ended after the December 1979 Soviet invasion.

Following the Soviet invasion, the United States supported diplomatic efforts to achieve a Soviet withdrawal. In addition, generous U.S. contributions to the refugee program in Pakistan played a major part in efforts to assist Afghans in need. U.S. efforts also included helping Afghans living inside Afghanistan. This cross-border humanitarian assistance program aimed at increasing Afghan self-sufficiency and helping Afghans resist Soviet attempts to drive civilians out of the rebel-dominated countryside. During the period of Soviet occupation of Afghanistan, the U.S. provided about $3 billion in military and economic assistance to Afghans and the resistance movement.

The U.S. Embassy in Kabul was closed in January 1989 for security reasons, but officially reopened as an Embassy on January 17, 2002. Throughout Afghanistan's difficult and turbulent 23 years of conflict, the U.S. supported the peaceful emergence of a broad-based government representative of all Afghans and actively encouraged a UN role in the national reconciliation process in Afghanistan.

Today, the U.S. is assisting the Afghan people as they rebuild their country and establish a representative government that contributes to regional stability, is market friendly, and respects human rights. The U.S. and Afghanistan are also working together to ensure that Afghanistan never again becomes a haven for terrorists. The U.S. provides financial aid for mine-clearing, reconstruction, and humanitarian assistance through international organizations.

Sources: U.S. Department of State (2004)
http://www.state.gov

Chapter References

Gyatso, Tenzin. 1990. *Freedom in Exile: The Autobiography of the Dalai Lama*. New York: HarperCollins.

Hedges, Chris. 1997. "Fascists Reborn as Croatia's Founding Fathers." *The New York Times* (April 12): Y3.

Jenkins, Emyl. 1996. *The Book of American Traditions*. New York: Crown Publishers

Lorimer, Lawrence. 1989. "Mennonite Churches." P. 213 in *The Universal Almanac 1990*, edited by J. W. Wright. Kansas City, MO: Universal Press Syndicate.

McGuire, Merridith B. 1987. *Religion: The Social Context*, 2nd ed. Belmont, CA: Wadsworth.

Steinfels, Peter. 1993. "Papal Birth-Control Letter Retains Its Grip." *The New York Times* (August 1):Y1+.

Yoachum, Susan and David Tuller. 1993. "Think Tank Tries to Prove Bible is Literal Truth." *San Francisco Chronicle* (September 14):A7.

Answers

Concept Application

1.	Ritual; Mystical religion	pg. 526
2.	Church, rituals	pg. 527; 526
3.	Liberation theology	pg. 545
4.	Sects	pg. 529
5.	Civil religion	pg. 532

Multiple-Choice

1.	c	pg. 521	13.	d	pg. 546
2.	b	pg. 522	14.	a	pg. 534
3.	b	pg. 524	15.	b	pg. 534
4.	b	pg. 525	16.	c	pg. 537
5.	b	pg. 526	17.	c	pg. 541
6.	c	pg. 526	18.	a	pg. 541
7.	c	pg. 524	19.	b	pg. 541
8.	d	pg. 527	20.	d	pg. 545
9.	a	pg. 529	21.	c	pg. 548
10.	d	pg. 530	22.	c	pg. 548
11.	d	pg. 532	23.	d	pg. 534
12.	d	pg. 534	24.	d	pg. 552

True/False

1.	F	pg. 520
2.	T	pg. 522
3.	T	pg. 526
4.	T	pg. 529
5.	T	pg. 532
6.	T	pg. 541
7.	T	pg. 543
8.	F	pg. 548
9.	F	pg. 550
10.	F	pg. 542

Chapter 16

Social Change
With Emphasis on Global Interdependence

Study Questions

1. How are the events of September 11 related to global interdependence?

2. Distinguish between global interdependence and globalization.

3. What is social change? Why is it an important topic within the discipline of sociology?

4. How do sociologists study change? What questions do they ask about social change?

5. Why is it difficult to predict the effects of a specific change?

6. What is an innovation? Distinguish between basic and improving innovations. What makes an innovation sociologically significant?

7. What is the cultural base? How is the rate of change tied to the size of the cultural base?

8. What is the information explosion? What technological innovations are responsible for this phenomenon? Explain.

9. What factors does Orrin Klapp identify as the causes underlying distorted, exaggerated presentation of information?

10. How is success of the internet related to efforts to control access to information?

11. What is cultural lag? Why did Ogburn emphasize the material component of culture in his theory of cultural lag?

12. Is Ogburn a technological determinist? Why or why not?

13. Ogburn maintains that one of the most urgent challenges facing people today is adapting to material innovations. Does the work of Leslie White lend support to Ogburn's thesis? Why or why not?

14. How does Kuhn define a paradigm?

15. According to Thomas Kuhn, is science simply an evolutionary process? Why or why not? Under what conditions are paradigms threatened? When does a scientific revolution occur?

16. How is conflict both a cause and an effect of social change?

17. Describe the essential dynamics of the Cold War and how those dynamics are connected with the development of the internet.

18. From a world system perspective, how has capitalism come to dominate the global network of economic relationships?

19. What is a social movement? What conditions are necessary for social movements to occur?

20. What are the types of social movements? Give a brief description of each.

21. Distinguish between objective and relative deprivation. How are these concepts related to social movements?

22. What are the three stages in the life of a social movement?

23. What is terrorism? Can governments engage in terrorist activities? Give an example.

Key Concepts

Global interdependence	pg. 562
Globalization	pg. 562
Social change	pg. 562
Innovation	pg. 566
Basic innovation	pg. 566
Improving innovation	pg. 566
Hypertext	pg. 569
Dearth of feedback	pg. 570
Cultural base	pg. 571
Invention	pg. 571
Simultaneous independent invention	pg. 571
Adaptive culture	pg. 572
Cultural lag	pg. 572
Technological determinist	pg. 572
Paradigm	pg. 572
Anomaly	pg. 573
Globalization from above	pg. 575

Globalization from below	pg. 575
Social movement	pg. 580
Regressive or reactionary movements	pg. 580
Reformist movements	pg. 581
Revolutionary movements	pg. 581
Counter revolutionary movements	pg. 582
Objective deprivation	pg. 582
Relative deprivation	pg. 582
Resource mobilization	pg. 583
Terrorism	pg. 583

Concept Application

Consider the concepts listed below. Match one or more of the concepts with each scenario. Explain your choices.

- ✓ Paradigms
- ✓ Hypertext
- ✓ Scientific revolution
- ✓ Technological determinism
- ✓ Globalization-from-above

Scenario 1 "It is difficult to recapture the medical world of 1800; it was a world of thought structured about assumptions so fundamental that they were only occasionally articulated as such, yet assumptions alien to a twentieth-century medical understanding.... The body was seen as a system of intake and outgo, a system that had to remain in balance if the individual were to remain healthy.... Equilibrium was synonymous with health, disequilibrium with illness.... The physician's most effective weapon was his ability to 'regulate the secretions' to extract blood, to promote the perspiration, the urination, or defecation that attested to his having helped the body regain its customary equilibrium" (Rosenberg 1987:71-72).

Scenario 2 Of course, Federal Express is our largest business unit by far. It is quite simply, the largest global express transportation network ever assembled. On our first night of operations - back in April of 1973 - we delivered just 186 packages to 25 U.S. cities, using a fleet of 14 small Falcon jets.

Twenty-five years later, FedEx delivers about 3 million shipments every business day to 211 countries that generate better than 90 percent of the world's GDP. The Federal Express workforce has grown to about 142,000 employees.... FedEx has the largest commercial cargo fleet in the world, with 615 aircraft, and about 100 more on order. That ranks us as the fourth largest airline worldwide - not just in the cargo industry, but among passenger airlines as well. In addition, the FedEx ground network includes about 42,000 trucks and vans, which are linked back into our data network to provide real-time information, from pick-up to delivery. And by utilizing one of the largest interactive computer networks in the world, better than 60 percent of all FedEx transactions are now handled electronically (Smith 1998).

Scenario 3 In public discussions of biotechnology today, the idea of improving the human race by artificial means is widely condemned. The idea is repugnant because it conjures up visions of Nazi doctors sterilizing Jews and killing defective children. There are many good reasons for condemning enforced sterilization and euthanasia. But the artificial improvement of human beings will come, one way or another, whether we like it or not, as soon as the progress of biological understanding makes it possible. When people are offered technical means to improve themselves and their children, no matter what they conceive improvement to mean, the offer will be accepted. Improvement may mean better health, longer life, a more cheerful disposition, a stronger heart, a smarter brain, the ability to earn more money as a rock star or baseball player or business executive. The technology of improvement may be hindered or delayed by regulation, but it cannot be permanently suppressed. Human improvement, like abortion today, will be officially disapproved, legally discouraged, or forbidden, but widely practiced. It will be seen by millions of citizens as a liberation from past constraints and injustices. Their freedom to choose cannot be permanently denied. (Dyson 1997:49)

Scenario 4 "Thomas Kuhn's seminal work, *The Structure of Scientific Revolutions*, affected working scientists as deeply as it moved those scholars who scrutinize what we do. Before Kuhn, most scientists followed the place-a-stone-in-the-bright-temple-of-knowledge tradition, and would have told you that they hoped, above all, to lay many of the bricks, perhaps even the keystone, of truth's temple, the addictive or meliorists model of scientific progress. Now most scientists of vision hope to forment revolution" (Gould 1987, p. 27).

Applied Research

Browse the internet to find examples of websites that facilitate globalization-from-above and globalization-from-below.

Practice Test: Multiple-Choice Questions

1. When studying a social change, sociologists ask
   a. Is social change good for society?
   b. How can we stop social change?
   c. Is social change necessary?
   d. What are the causes and consequences of change for social life?

2. _________________ is any significant alteration, modification, or transformation in the organization and operation of social life.
   a. Globalization
   b. Scientific revolution
   c. Social change
   d. Global interdependence

3. Each upgrade of a personal computer's CPU (central processing unit) represents a
   a. discovery.
   b. a basic innovation.
   c. an improving innovation.
   d. a paradigm shift.

4. Optical fibers have the potential to transmit in a few seconds time the equivalent of
   a. 1200 letters of the alphabet.
   b. a large telephone directory.
   c. a 1300-page novel.
   d. the entire contents of the Library of Congress.

5. Which analogy did sociologist Orrin Klapp use to describe the dilemma of sorting through and keeping up with the massive amount of information being generated?
   a. A sociologist drowning in quicksand.
   b. A student trying to take notes while 10 professors talk at one time.
   c. A researcher working on a gigantic jigsaw puzzle while additional pieces are flowing onto the table from a funnel overhead.
   d. A person entering a crowded six-lane highway with thousands of signs.

6. The U.S. Environmental Protection Agency confronted the question of whether to place on the internet "worst case" scenario data to help the public plan and prepare for potential chemical accidents. In the end the EPA decided to
   a. post the information on the internet.
   b. withhold the information from the public.
   c. place information in city buildings across the U.S.

   d. make the information available in 50 federal reading rooms across the country.

7. "We invent the automobile to get us between two points faster, and suddenly we find we have to build new roads. And that means we have to invent traffic regulations... and then we have to invent a whole new organization called the highway patrol." This assessment supports the idea that
   a. necessity is the mother of invention.
   b. if a new invention is to come into being, the cultural base must be large enough to support it.
   c. invention is the mother of necessity.
   d. if people have the power to create material innovations they also have the power to destroy them.

8. ________________ are situations in which the same invention is created by two or more people working independently of one another at about the same time.
   a. Cultural diffusions
   b. Scientific revolutions
   c. Improving inventions
   d. Simultaneous-independent inventions

9. The explanatory value and hence the status of a paradigm is threatened by the existence of an anomaly. An anomaly is
   a. a dominant and widely accepted theory.
   b. an observation that the paradigm cannot explain.
   c. a modification of a basic invention.
   d. a transformation of the social structure.

10. When a new paradigm causes converts to see the world in an entirely new light and wonder how they could possibly have taken the old paradigm seriously, _______ has occurred.
    a. a scientific revolution
    b. innovation
    c. cultural lag
    d. adaptive reasoning

11. Perhaps the most outstanding feature of the internet is that it was designed to operate
    a. from a central command station in Washington.
    b. on solar power.
    c. automatically.
    d. without a central control over it.

12. The symbolic end to the Cold War came when
    a. the coup against Mikhail Gorbachev failed.
    b. the Berlin Wall was dismantled.
    c. West and Easter Germany reunified.
    d. the Soviet Parliament voted to dismantle the Communist Party.

13. Marx believed that _________ was the first economic system capable of maximizing the immense productive potential of human labor and ingenuity.
    a. the capitalist system
    b. socialism
    c. communism
    d. a centrally planned economy

14. Most UPS employees are based in
    a. Europe.
    b. Japan.
    c. the United States.
    d. the Middle East.

15. A social movement depends on three conditions. Which one of the following is <u>not</u> one of those conditions?
    a. an actual or imagined condition that enough people find objectionable
    b. a shared belief that something needs to be done about this condition
    c. some organized effort aimed at attracting supporters, articulating the problem, and defining a strategy.
    d. enough financial support to get the movement off the ground.

16. Befrienders International is a ____________ movement that seeks to prevent suicide.
    a. regressive
    b. reformist
    c. revolutionary
    d. counter revolutionary

17. Tyrone earns an annual income of $100,000. His friends earn between $300,000 and $500,000 a year. Tyrone feels left out because he cannot afford the kind of cars his friends drive. Tyrone is experiencing
    a. objective deprivation.
    b. relative deprivation.
    c. regressive thoughts.
    d. a feeling of being an anomaly.

18. According to sociologist Ralf Dahrendorf, the structural origins of conflict can be traced to
    a. the nature of authority relations.
    b. decision-making powers of the power elite.
    c. invention and innovation.
    d. workers' demands for higher wages.

19. Ralf Dahrendorf wrote "It is immeasurably difficult to trace the path on which a person...encounters other people just like himself, and at a certain point says 'Let us join hands, friends, so that no-one will push us off one by one.'" Dahrendorf was writing about
    a. globalization-from-above.
    b. social movements.
    c. globalization-from-below.
    d. the Chernobyl meltdown.

True/False Questions

T F 1. Sociology first emerged as a discipline attempting to understand social change.

T F 2. Globalization emerged with the invention of the internet.

T F 3. Geometric expansion can be represented by the following sequence: 1, 2, 4, 8, 16, 32...

T F 4. From a sociological point of view, invention is the mother of necessity.

T F 5. Some inventions, such as the bicycle, generate no conflict in society.

T F 6. The internet is an example of a technology that emerged out of cooperation between the U.S. and the former Soviet Union.

T F 7. Capitalist responses to economic stagnation and downturn helped to create a global network of economic relationships.

T F 8. Research on social movements shows that the most objectively disadvantaged people join social movements to change their condition.

T F 9. Each year about one dozen terrorist attacks take place on U.S. soil.

T F 10. The structural source of conflict lies with formal authority structures.

T F 11. Terrorist attacks tend to target those directly responsible for alleged grievances.

Internet Resources Related to Globalization and Global Interdependence

- **The World Bank Group: Globalization**
  http://www.worldbank.org/economicpolicy/globalization
  The World Bank website offers an overview of globalization as it relates to poverty and a world economy. The website offers a list of key readings on the topic as well as data and statistics.

- **The Globalization Website**
  http://www.emory.edu/soc/globalization
  The website offers an overview of the concept globalization and addresses six key questions: "(1) What is globalization? (2) How does globalization affect women? (3) Does globalization cause poverty? (4) Why are so many people opposed to globalization? (5) Does globalization diminish cultural diversity? and (6) Can globalization be controlled?"

Statistical Profile: International Arrivals to the U.S. by Country of Residence

Globalization is the ever increasing "cross border flows of goods, services, money, people, information, and culture" (Held, McGraw, Goldblatt, and Perraton 1999). One example of globalization relates to international tourism which was estimated at 50,000 arrivals in 1950 and reached 666.9 million in 1999. The table below shows the international arrivals to the U.S. by country of residence in 2000.

Country of Origin	Number of arrivals to U.S. 2000
Albania	5,737
Algeria	3,294
Andorra	1,235

Angola	3,146
Anguilla	5,177
Antigua & Barbuda	24,918
Argentina	533,936
Armenia	8,663
Australia	539,559
Austria	175,533
Azerbaijan	1,951
Bahamas	293,911
Bahrain	6,735
Bangladesh	12,724
Barbados	57,071
Belgium	249,957
Belize	30,352
Bermuda	7,474
Bolivia	49,135
Bosnia and Herzegovina	5,430
Botswana	2,391
Brazil	737,245
British Virgin Islands	29,708
Brunei	1,230
Bulgaria	11,418
Burma	1,357
Byelarus	4,444
Cameroon	3,124
Cape Verde	3,906
Cayman Islands	52,922
Canada 1	14,594,000
Canada Air Only 1	5,300,000
Chile	192,361
China (PRC) / Hong Kong 2	452,741
China (PRC) 3	249,441
Hong Kong	203,300
Colombia	417,065
Congo	1,604
Costa Rica	176,056
Croatia	11,244
Cuba	48,614
Cyprus	12,152
Czech Republic	43,758
Czechoslovakia	11,590
Denmark	149,211
Dominica	16,174
Dominican Republic	197,298
Ecuador	129,938
Egypt	44,612
El Salvador	184,574

Estonia	6,745
Ethiopia	5,177
Fiji	6,817
Finland	93,649
France	1,087,087
French Polynesia	3,373
Gabon	1,404
Gambia	2,966
Georgia	2,173
Germany	1,786,045
Ghana	16,073
Gibraltar	1,363
Greece	61,361
Grenada	10,412
Guadeloupe	10,253
Guatemala	185,677
Guinea	4,184
Guyana	17,435
Haiti	72,190
Honduras	90,714
Hungary	59,174
Iceland	27,682
India	274,202
Indonesia	71,390
Iran	9,364
Ireland	285,697
Israel	325,199
Italy	612,357
Ivory Coast	5,226
Jamaica	242,903
Japan	5,061,377
Jordan	17,327
Kampuchea	2,319
Kazakhstan	3,311
Kenya	14,229
Republic of Korea	661,844
Kuwait	23,930
Latvia	7,161
Laos	1,045
Lebanon	18,599
Liberia	3,823
Liechtenstein	2,011
Lithuania	9,122
Luxembourg	16,385
Macau	4,077
Macedonia	2,841
Malawi	1,365

Malaysia	74,507
Mali	4,015
Malta	8,437
Martinique	9,912
Mauritius	1,899
Mexico 1	10,322,000
Mexico (Air Only)	1,736,609
Moldova	1,625
Monaco	5,622
Montserrat	1,027
Morocco	19,079
Namibia	1,144
Nepal	4,101
Netherlands	553,297
Netherlands Antilles 4	60,042
New Caledonia	1,759
New Zealand	172,012
Nicaragua	46,892
Niger	1,433
Nigeria	27,124
Norway	147,540
Oman	5,396
Pakistan	51,942
Panama	107,349
Papua New Guinea	1,336
Paraguay	18,714
Peru	192,062
Philippines	168,053
Poland	116,277
Portugal	86,333
Qatar	5,860
Romania	27,418
Russia	76,739
Saudi Arabia	75,320
Senegal	11,185
Singapore	136,439
Slovakia	13,235
Slovenia	14,886
South Africa	116,113
Spain	361,177
Sri Lanka	9,898
St. Kitts-Nevis	12,084
St. Lucia	18,100
St. Vincent-Grenadines	7,291
Sudan	1,479
Suriname	7,039
Sweden	321,881

Switzerland	395,031
Syria	8,470
Taiwan	457,302
Tanzania	4,286
Thailand	86,971
Togo	1,590
Tonga	2,719
Trinidad & Tobago	137,689
Tunisia	9,489
Turkey	106,427
Turks and Caicos Islands	16,127
Uganda	3,885
Ukraine	14,477
United Arab Emirates	40,039
United Kingdom	4,703,008
Uruguay	69,607
Uzbekistan	6,809
Venezuela	576,663
Vietnam	9,061
Western Samoa	1,206
Yemen	1,995
Yugoslavia	8,039
Zambia	3,337
Zimbabwe	8,080

Chapter References

Dyson, Freeman. 1997. "Can Science Be Ethical?" *The New York Review* (April 10):46.

Goldstein, Steven M. 1991. *Minidragons: Fragile Economic Miracles in the Pacific*. New York: Ambrose Video.

Gould, Stephen J. 1987. *An Urchin in the Storm: Essays about Books and Ideas*. New York: Norton.

Rohter, Larry. 1997. "Trade Storm Imperils Caribbean Banana Crops." *The New York Times* (May 9):A6.

Rosenberg, Charles E. 1987. *The Care of Strangers: The Rise of America's Hospital System*. New York: Basic Books.

Smith, Frederick W. 1998. "Defining the global economy." *Vital Speeches*. (Dec 1):v65 i4 p125(4).

Answers

Concept Application

1.	Paradigms	pg. 572
2.	Globalization from above	pg. 575
3.	Technological determinism	pg. 572
4.	Scientific revolution, paradigm	pg. 572

Multiple-Choice

1.	d	pg. 567
2.	c	pg. 562
3.	c	pg. 566
4.	d	pg. 569
5.	c	pg. 569
6.	d	pg. 570
7.	c	pg. 571
8.	d	pg. 571
9.	b	pg. 573
10.	a	pg. 573
11.	d	pg. 584
12.	b	pg. 574
13	a	pg. 575
14.	c	pg. 579
15.	d	pg. 580
16.	b	pg. 581
17.	b	pg. 582
18.	a	pg. 582
19.	b	pg. 583

True/False

1.	T	pg. 562
2.	F	pg. 562
3.	T	pg. 571
4.	T	pg. 571
5.	F	pg. 573
6.	F	pg. 574
7.	T	pg. 578
8.	F	pg. 582
9.	F	pg. 574
10.	T	pg. 582
11.	F	pg. 584